Epoxy Resin Chemistry II

Epoxy Resin Chemistry II

Ronald S. Bauer, EDITOR
Shell Development Company

Based on a symposium sponsored
by the ACS Division
of Organic Coatings and
Plastics Chemistry
at the 183rd Meeting
of the American Chemical Society,
Las Vegas, Nevada,
March 28–April 2, 1982

ACS SYMPOSIUM SERIES 221

AMERICAN CHEMICAL SOCIETY
WASHINGTON, D.C. 1983

Library of Congress Cataloging in Publication Data

Epoxy resin chemistry II.
(ACS symposium series, ISSN 0097-6156; 221)

Includes bibliographies and index.

1. Epoxy resins—Congresses. I. Bauer, Ronald S., 1932- . II. American Chemical Society. Division of Organic Coatings and Plastics Chemistry. III. Title: Epoxy resin chemistry 2. IV. Series.

TP1180.E6E588 1983 668'.374 83-6385
ISBN 0-8412-0777-1

PRINTED IN THE UNITED STATES OF AMERICA

ACS Symposium Series

M. Joan Comstock, *Series Editor*

FOREWORD

The ACS SYMPOSIUM SERIES was founded in 1974 to provide a medium for publishing symposia quickly in book form. The format of the Series parallels that of the continuing ADVANCES IN CHEMISTRY SERIES except that in order to save time the papers are not typeset but are reproduced as they are submitted by the authors in camera-ready form. Papers are reviewed under the supervision of the Editors with the assistance of the Series Advisory Board and are selected to maintain the integrity of the symposia; however, verbatim reproductions of previously published papers are not accepted. Both reviews and reports of research are acceptable since symposia may embrace both types of presentation.

CONTENTS

PREFACE

NOT MANY SYNTHETIC ORGANIC MATERIALS rival epoxy resins in their diversity of application. The use of epoxy resins in applications such as protective coatings, adhesives, electrical laminates, reinforced plastics, and commercial flooring systems is a testimony to the versatility and high performance characteristics of these materials. Although epoxy resin sales have not achieved the volumes that commodity-type thermoplastic materials have, their growth has been impressive. Introduced commercially in the United States in the late 1940s, epoxy resins achieved annual sales of about 13 million pounds by 1954. In 1981, total demand was about 319 million pounds, and that figure is expected to grow to 510 million pounds in 1990.

Future demands for epoxy resins will depend largely on the results of research such as described in this volume, as well as the outcome of future efforts. As judged by the number of publications appearing each year on epoxy resins, the research effort devoted to these materials has not shown any sign of easing. A survey of the number of publications appearing per year between 1971 and 1981 that refer to epoxy resins and that are cited in "Chemical Abstracts" indicates that the number has remained relatively constant (*see* Table), at between 2200 and 2300.

The symposium on which this volume is based was an attempt to represent a cross-section of the current research being conducted on epoxy resins. Included is work on efforts to improve the performance of existing products through rubber modifications that will expand markets for resins and innovative products based on currently available commercial resins. Also discussed is the work in progress to develop a better understanding of the properties of cured systems so as to obtain from them their ultimate performance. If the research effort of the last decade is sustained through the next decade, the new innovations in epoxy resins could create a demand for resin that would easily exceed current expectations. When the ingenuity of chemists and engineers is coupled with the versatility and high performance characteristics of epoxy resins, the result will be the development of a multitude of new end-uses and currently unanticipated markets.

Number of Epoxy Resin Publications Cited in "Chemical Abstracts" Between 1971 and 1981

Year of Publication	*Patents*	*Nonpatents*	*Total*
1981	1417	824	2241
1980	1944	910	2314
1979	1198	1108	2306
1978	1165	1087	2252
1977	1131	1073	2204
1976	1273	889	2162
1975	1399	967	2366
1974	1320	861	2181
1973	1109	840	1949
1972	941	144	1785
1971	684	737	1421

RONALD S. BAUER
Shell Development Company
Houston, Texas 77001

February 1982

1

Elastomer-Modified Epoxy Resins in Coatings Applications

R. S. DRAKE and D. R. EGAN

B F Goodrich Chemical Group, Cleveland, OH 44131

W. T. MURPHY

B F Goodrich Company, Research and Development Center, Brecksville, OH 44141

Elastomer-modified epoxy resins have grown in usage in the past 15 years in several application areas connected with structural adhesives, composites, civil engineering/construction, electrical laminates/ encapsulants and corrosion resistance. Some of this increased attention has come about through utility of telechelic butadiene/acrylonitrile liquid polymers. Both carboxyl and amine reactive liquid polymers (CTBN and ATBN) have provided chemistries amenable to this modification with the polybutadiene/acrylonitrile copolymer providing solubility parameters close to if not equaling those of base epoxy resins. It has only been within the last few years that similar elastomer-modified epoxy resins have been examined in epoxy coatings and primers. This chapter reviews a portion of that relatively new journal and patent literature involving epoxy coal-tar, powder, photo-curable and solventless heavy duty coatings as well as metal primers. These examples are employed to illustrate such benefits of elastomer inclusion as reverse impact, bending/crimping, corrosion resistance, thermal shock resistance and coating adhesion. The chapter concludes with an initial reporting on industrial maintenance and marine coatings based on elastomer-modified epoxy resin models. Similar benefits are noted with these traditional solvent-based coatings under ambient cure.

Numerous blends or alloys of thermosetting resins with elastomers have been developed over the past four decades and found their way into continuing commercial use. A generous amount of this has been related to toughening needs: Overcoming the brittleness of highly crosslinked glassy polymers. With it came the recognition that not only was degree of elastomer/resin compatibility important but also the need for achieving a dispersed phase

0097-6156/83/0221-0001$06.00/0

with particle size integrity. Thus, both blended or alloyed products, as well as the processing techniques for achieving them, were optimized and later improved upon. Products for both structural and non-structural applications proliferated.

Among the earlier developments of elastomer/resin blends, the polybutadiene/acrylonitrile or 'nitrile' elastomer modification of phenol-formaldehyde condensation resins serves as a typical example and a pivot for this paper's thesis as well. The nitrile-phenolic category of products attests to a large and continuing literature showing first the usage of nitrile elastomers themselves and later their carboxylic analogs. Systematic evaluations of elastic phenol-formaldehyde molding compositions, e.g., with both carboxylic and analogous noncarboxylic elastomers are useful in demonstrating the stronger reinforcement properties of the carboxyl-reactive species.

It is these solid carboxylic nitrile elastomers which began to show utility in the modification of epoxy resins. Processing needs for solid elastomer inclusion, particularly in liquid epoxy resins, have not always been advantageous. Associated problems include gel, viscosity threshold limitations and achieving desired rubber levels in excess of 5-6 phr. Sometimes processing must be carried out in selected solvents, not always a desirable or tolerable step.

In the mid-60's carboxyl-terminated polybutadiene/acrylonitrile (CTBN) liquid polymers were introduced for the purpose of epoxy resin modification. These telechelic polymers are essentially macromolecular diacids. They offer processing ease (and therefore advantage) over the solid carboxylic nitrile elastomers. It is no surprise that the epoxy prepreg industry (adhesive and non-adhesive varieties) found the liquid and solid carboxylic nitrile elastomer species useful together in processing liquid and lower molecular weight solid epoxy resins where elastomer modification was needed.

Later, in 1974, amine reactive versions of the liquid nitrile polymers (ATBN) were issued, thereby offering another way to introduce rubbery segments into a cured epoxy resin network. References are cited which provide detailed discussions of nitrile rubber, carboxylic nitrile rubber and both carboxyl- and amine-terminated nitrile liquid polymers (1-4). Table I illustrates CTBN and ATBN products structurally. Table II provides properties for typical solid carboxylic nitrile elastomers.

It is the purpose of this review paper to examine the relatively newer uses of these elastomer types as resin modifiers within the broad range of epoxy-based coating systems.

Elastomer-Modified Epoxy Resin Preparation

Wide-ranging documentation exists which covers modification of epoxy resins using carboxyl-terminated polybutadiene/acrylonitrile liquid polymers in which addition esterification (alky-

TABLE I

CARBOXYL-TERMINATED (CT) LIQUID POLYMERS....

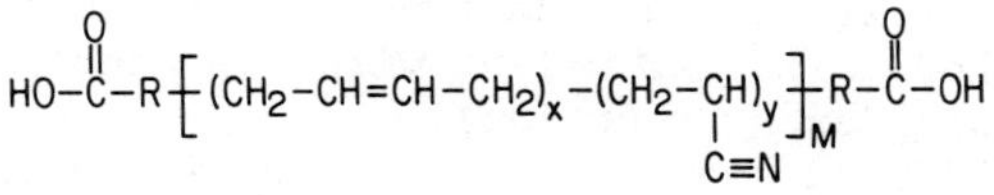

Carboxyl Equivalent Weight (Typical) = 1850

AMINE-TERMINATED (AT) LIQUID POLYMERS ....

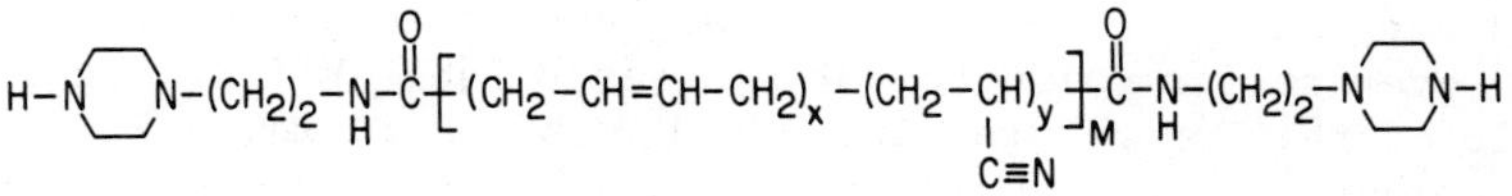

Amine Equivalent Weight (Typical) = 900

TABLE II

SOLID CARBOXYLIC BUTADIENE/ACRYLONITRILE ELASTOMERS

Physical Form	Percent Acrylonitrile	EPHR Carboxyl	Mooney Viscosity (ML-4' @ 100°C)
Crumb*	26-28	0.07-0.08	35-45

*Hycar 1472, a product of BF Goodrich Chemical Group

droxyl esterification) reactions are employed to prepare the epoxy base (5-8). While such reactions proceed uncatalyzed at excess epoxy equivalents normally greater than 3/1-epoxy/carboxyl, instructive examples are given here for t-phosphine or quaternary phosphonium salt catalyzed systems. Table III illustrates this for a common liquid diglycidyl ether of bisphenol A (DGEBA) at three selected temperatures with time-to-zero-acid measured. Table IV shows a variation on this with mixed diacids of varying pKa value employed. Tables V and VI give recipes for the preparation of elastomer-modified solid DGEBA epoxy resins either using a one-step method (direct modification) or a two-step method (advancement to higher molecular weight with bisphenol A after completing the carboxyl/epoxy reaction).

Elastomer-modified epoxy resin systems with more complexity to their preparation scheme have been demonstrated. Two examples suffice. Shelley and Clarke (9) instruct that a 'vulcanization' procedure can be successfully employed to improve elevated temperature properties in the cured resin mass. This step occurs subsequent to the esterification regime. It can be practiced with impunity at low rubber contents (7.5-10%) without gelation or indeed very much viscosity increase. Peroxides appear to be preferred over sulfur/sulfur donor systems. Table VII displays an example of this procedure with a solid DGEBA resin.

Another chosen example stems from the perceived benefits of combining solid and liquid carboxylic nitrile elastomers in the same modification scheme. This is particularly useful when a degree of 'tack' is required in the system. Table VIII exhibits one approach for preparing such an elastomer-modified solid DGEBA resin. In this instance a resin (epoxide equivalent weight of 650) was prepared by advancing two available liquid epoxy resin adducts -- one utilizing a liquid nitrile, the other a solid nitrile elastomer.

To a lesser extent, amine-terminated polybutadiene/acrylonitrile (ATBN) liquids are also used by epoxy formulators. This polymeric disecondary amine is employed in admixture with amines, amidoamines or fatty polyamides of choice (10, 11). Thus, one admixes chemical and polymeric amines to create a novel toughening or flexibilizing hardener package.

In each instance of nitrile elastomer modification - whether rubber is added to the epoxy portion or to the hardener portion - the level of rubber largely determines whether a toughened or a flexibilized epoxy results. The former is characterized by little loss in thermal/mechanical properties. The latter shows a dominant influence of the added rubber.

Consequent to documentation surrounding methods of employing reactive nitrile elastomers to modify epoxy resins is a growing body of literature which serves to characterize and elucidate these systems. Such topics as morphology in the cured and uncured state, transitions from toughening to flexibilization, viscoelastic effects, equilibrium physical properties and phase structure are available to the investigator (12-17).

TABLE III

ADDUCTS OF LIQUID EPOXY AND CTBN

	Parts (Weight)
DGEBA Liquid Epoxy (EEW, 185)	62.50
CTBN (1300X8)	37.50
Triphenylphosphine	0.25

Temperature (°C)	Time to Zero Acid (Hrs)
60	6.0
80	1.3
120	<0.5

TABLE IV

ADDUCTS OF LIQUID EPOXY, CTBN AND DIMER ACID

	Parts (Weight)
DGEBA Liquid Epoxy (EEW, 185)	75
Dimer Acid (Acid Value, 192)	15
CTBN (1300X8)	10
Triphenylphosphine	0.2

T/t - 165°C/45 Mins.

EEW - 275-285
B.V. mPa·s at 27°C - 36,000-42,000

TABLE V

SOLID EPOXY RESINS WITH CTBN
ONE-STEP APPROACH

	Parts (Weight)
DGEBA Liquid Epoxy (EEW, 185)	100
Bisphenol A	24
CTBN	variable
Triphenylphosphine	0.2

T/t 165-175°C/1 Hr.

TABLE VI

SOLID EPOXY RESINS WITH CTBN
TWO-STEP APPROACH

1. Prepare CTBN (1300X13)/DGEBA Epoxy (EEW, 185) at 40/60 (See Table III)

		Parts (Weight)
2.	Adduct (1)	100
	Bisphenol A	16
	Ethyltriphenylphosphonium Iodide	0.08

T/t - 110°C/6 Hrs.
EEW Actual - 743
EEW Theoretical - 758
Rubber Content - 35%
Form - Semi-solid

TABLE VII

SOLID EPOXY RESINS WITH CTBN
USING 'VULCANIZATION' STEP

Heat to 100°C	Parts (Weight)
DGEBA Liquid Epoxy (EEW, 185)	100
Bisphenol A	28.7
CTBN (1300X8)	10
Add	
Catalyst - Tetrabutyl Phosphonium Acetate acetic acid complex (as 7% solution in methanol)	0.20
T/t - 140°C/30 Mins.	
Add	
Dicumyl Peroxide	0.33
T/t - 160°C/2.5 Hrs.	

EEW - 555
Durran's S.P. - 70°C

Ref: Shelley, Clarke (BP 1,461,127)

TABLE VIII

SOLID EPOXY RESINS WITH LIQUID AND SOLID CARBOXYLIC ELASTOMERS

	Parts (Weight)
WC-901A (EEW, 210)[1]	90
Kelpoxy 272-100 (EEW, 340)[2]	25
Bisphenol A	31.5
Ethyltriphenylphosphonium Iodide	0.15

T/t - 110°C/6 Hrs.
EEW - 648
Rubber Content - 10.2 phr

[1]A product of Wilmington Chemical Co.
[2]A product of Spencer-Kellogg, Inc.

Elastomer-Modified Epoxy Resin Applications

Nitrile-rubber toughened epoxy resins for formulating structural adhesives and fiber-reinforced composite matrix resins was reviewed in 1975 (18). More recently, Bascom et al (19) have shown that the generous increases obtainable in bulk fracture energy (due to rubber inclusion) are indeed transferred to a comparable interlaminar fracture toughness in the composite (woven glass and woven graphite as models). Following this, Moulton and co-workers (20) described the coming generation of high performance advanced composites. In each instance rubber modification has been used to benefit.

The nitrile-epoxy structural adhesive systems have moved from an aircraft orientation to include industrial, automotive and electronics areas. These include a host of film adhesive products (21, 22) and thixotropic pastes (23-26). Representative adhesive models have also been studied from a fracture toughness point-of-view permitting comparisons of bulk fracture to that of fracture in the adhesive joint (27, 28).

Small nitrile-rubber inclusions in epoxy resin electrical encapsulants have been examined in both amine (29-31) and acid (32) epoxy cures, in filled and unfilled systems. The value of rubber inclusion in a boron trifluoride/amine complex epoxy cure has also been demonstrated (33), where elevated-temperature, high-humidity testing showed electrical properties retention to be better than a comparable system cured with dodecenylsuccinic anhydride. Rubber benefits low-temperature properties specifically and thermocycling in general. It affects high temperature insulation properties negatively; therefore, the amount of rubber incorporated must be judiciously chosen.

In the associated areas of epoxy corrosion-resistance, both carboxyl- and amine-reactive nitrile liquid rubbers are employed in filament-wound structures, pipe liners, castings and flooring systems. Rubber-modified epoxy-methacrylates may be prepared from the carboxyl-reactive polymers (34, 35) with specialty uses within the important corrosion-resistant fiber-reinforced plastics markets (36, 37). Amine-reactive nitrile liquid rubber is used with aromatic amines or their eutectics in curing epoxy resins where levels of rubber up to 10 phr have minimal effect in most chemical environments. At even higher amine-reactive nitrile liquid rubber levels, one finds liner systems (metal surfaces), flooring (concrete, masonry) and sealants (civil engineering/construction) where ATBN is employed in conjunction with chemically modified aromatic amines designed to cure epoxy resins under ambient conditions or with select cycloaliphatic amines such as isophorone diamine. Such applications for ATBN type products confirm its utility in formulating novel epoxy hardeners.

It is these corrosion-resistant or acid-resistant uses for

nitrile elastomer modified epoxy resins which have influenced the more recent coating and primer developments.

Elastomer-Modified Epoxy Resins - Coatings Background

Siebert and Riew (38) as early as 1971 described a CTBN-toughened epoxy resin model (piperidine cured) applied and tested as a coating on cold rolled steel. The appearance of a uniformly dispersed elastomeric second phase was identified and bulk fracture toughness measured (20 fold improvement) with elongations increasing from 4.8 to 6.2% at 10 phr rubber. Bending/crimping tests (T-Blend-3/16 inch radius) also showed a practical advantage to elastomer inclusion in terms of retained adhesion.

Later, McPherson (39) demonstrated that the inclusion of a CTBN/liquid epoxy adduct enhanced both the flexibility and adhesion of a dimer acid flexibilized epoxy/brominated epoxy mixture when cured with a combination of chlorendic and nadic-methyl anhydrides. With an epoxy-glass laminate as substrate, such mixtures were coated at 0.5-2.0 mils thickness, cured, sensitized and copper plated (1.5 mils) by an electroless process. 90° peel testing (2 inches/min. rate, 25°C) gave 14.0 pli peel force where 5.0 pli is considered normal.

From formulated flexibilized epoxy resin admixtures having improved adhesion with the use of CTBN/epoxy adducts, it was demonstrated that a novel epoxy elastomer (40) based on CTBN/epoxy adducts in combination with coal tar could be achieved. Cured under ambient conditions, this system demonstrated good elasticity retention at -10°C, 100°C and long-term ambient aging. Further testing suggested uses as impermeable coatings (such as the waterproofing) where water absorption tests (30 days-ISO R62) registered only 1.5%.

At lesser rubber levels, heavy duty solventless coatings models based on diethylenetriamine or fatty polyamide cures, showed elastomer-modification (10 phr rubber level) to advantage in Gardner impact, mandrel bend and corrosion-resistance testing (41). Impact testing (direct and reverse) gave 110 and 60 in-lbs, respectively for the rubber-modified fatty polyamide cured epoxy coating (14 days at R.T.), whereas a control formulation tested 10 in-lbs in each mode.

Lewis and co-workers (42) developed improved powder coatings with nitrile rubber-modification of an appropriate epoxy base (solid resin admixture) cured with an imidazoline-accelerated modified phenolic type hardener. Model coatings ground to 55 μm particle size, electrostatically applied to metals, cured 10' @ 170°C, gave excellent thermocycling results as well as retained resistance to solvent attack. Elastomer-modified epoxy powder coatings have been covered extensively by Geibel, Romanchick and Sohn in Chapter 5 of this volume.

Earlier, McKown (43) had shown that nitrile-rubber modified type solid epoxies (EEW ~900) could be prepared and applied successfully in particulate form under electrostatic conditions and used as high-peel structural adhesives with no loss in shear strength properties. Patent literature (44) continues to spell out unique flexible powder compositions where dicyandiamide cured epoxy resins accelerated with substituted imidazoles utilize elastomer-modified epoxy resins in conjunction with select thermoplastic additives.

Ariga (45) disclosed developments with CTBN-modified epoxy acrylates which when prepared with other lower molecular weight dicarboxylic acids and select carboxyl-terminated oligopolyesters serve as novel oligomers for photocurable coatings. These systems have low uncured viscosity. In the cured state they possess excellent scratch and crosshatch adhesion, good direct and reverse impacts (duPont impact tester) and pass bending diameters of 2 mm.

Diener (46) and Tsuchiya (47) have shown carboxyl-reactive nitrile liquids to have utility in both aqueous and non-aqueous anodic and cathodic electrodeposition systems aimed at primers and coatings. Excellent coating adhesion is demonstrated with advantages noted in moisture resistance and reverse impacts. Diener suggests an electrocoat system as a replacement for standard solvent-based primers used with aircraft adhesives.

One can rationalize a need for small rubber inclusions in some of the newer approaches to waterborne and high solids epoxy coating systems. Water-thinned epoxy coating compositions are described (48) where the two-component system consists of a nitrile rubber modified epoxy resin in the epoxide component and a styrene/butadiene/methylmethacrylate latex modifier for an emulsion-based polyamide hardener component. Showing improved adhesion, impact and water resistance, the paint has good wetting characteristics and can be formulated to a high solids content at low viscosity.

Recently, Athey (49) reviewed telechelic polymer synthesis and illustrates with examples that these materials may be used effectively in high solids coatings to chain extend or cure in situ to form appropriate coating vehicles.

Thicker epoxy-based coatings, highly flexibilized with amine-reactive nitrile liquid polymer, have been described by Mendelsohn (50) in which the flexibilizing hardener is comprised of an admixture of ATBN, fatty polyamide and boron trifluoride/amine complex. This coating ages well at 100°C and has excellent toughness/flex with good abrasion and vibration absorption properties. Heavy, filled construction coatings have been developed where ATBN admixtures with meta-xylylenediamine constitute an unique flexibilizing hardener for the lower viscosity liquid epoxy resin based on judicious mixtures of bisphenol A and bisphenol F (51).

Numata and Kinjo (52) have shown rubber-modified isocyanurate-oxazolidone resins may be effectively modified with carboxyl-reactive nitrile liquids. The viscoelastic behavior of models using a polyglycidyl ether of phenol-formaldehyde novolac resin and diphenylmethane-4,4'-diisocyanate is discussed. Such resins have suggested utility in thin films as electrical varnishes.

Ambient Temperature Cured Solution Epoxy Coatings

Conventional DGEBA type epoxy resins serve as the polymeric film-former for solventless, high solids, waterborne and solution coatings used for protection purposes. These resins are converted into useful thermoset coatings through the introduction of appropriate curing agents (53). This chapter concludes with an examination of elastomer-modified epoxy resin solution coating models under ambient cure.

These types of coatings are used to protect metals from the devastating loss attributed to corrosion, estimated to be greater than 20 billion dollars annually in the U.S. alone. More specifically, epoxy coatings are employed in numerous industrial applications requiring the protection of structural steel. Case histories of such application are prevalent in the literature (54). Representative of industries using solution epoxy coatings are industrial maintenance (storage tanks, plant equipment, etc.) and marine (hull, decks, etc.). Two-component coatings, allowing for ambient temperature curing, are extremely useful to these industries in which baking a protective finish is either impractical or impossible.

Two package epoxy coatings may be used as primer, intermediate and topcoat. The more visible coatings of this type consist of epoxy resin with amine or polyamide hardeners. The common epoxy resin is a DGEBA type having a molecular weight of approximately 1000. Film properties are, of course, determined largely by the hardener type used. Amines from the polyethylene homolog series such as diethylenetriamine (DETA) and triethylenetetramine (TETA) will cure an epoxy coating under ambient conditions to relatively high crosslink density. Unmodified polyfunctional aliphatic amines are seldom used for health reasons. To circumvent these problems, chemical modifications are carried out (55).

Higher molecular weight epoxy resin hardeners such as fatty polyamides do not have the problems associated with amines and amine adducts. These reaction products of dimer and trimer acids with polyfunctional amines provide films with fewer surface discontinuities. Polyamide cured epoxy coatings are more forgiving than amine cured epoxy coatings since they require less demanding surface preparation. Also, mix ratios are less critical for polyamide/epoxy coatings (56).

This study of elastomer-modified solution epoxy coatings uses typical epoxy-polyamide and epoxy-polyfunctional amine based models.

Elastomer Modification

Elastomer modification of two-package epoxy coatings has been carried out either with rubber addition to the epoxy component or to the hardener component. Amine-reactive nitrile liquids have been examined extensively in polyamide cured epoxy coatings. The actual model formulation consists of a DGEBA solid epoxy (epoxide equivalent weight of 450-550) and polyamide (amine value of 230-246). The model formulation, with data on clear and pigmented versions, appears in Tables IX and X. The substrate used in this study is cold rolled steel uniformly sandblasted to a specified profile.

The precedence for nitrile-epoxy alloys in other epoxy applications suggests that similar compositions in coatings should enhance film properties associated with flexibility, impact resistance and adhesion to metals. The testing scheme employed emphasizes these property features.

Dynamic-Mechanical Property Response

Rubber-toughened epoxy resins have been characterized by their resin and rubber glass transition temperatures (57) using torsional braid analysis (TBA). These are reported in the literature as eTg and rTg, respectively. The Torsional Braid Analyzer has also been used to characterize ATBN-modified coatings regarding transition temperatures.

Examining these model epoxy coatings for transition temperatures shows that resin Tg is maintained throughout the series of rubber modification. See Figure 1. Data points associated with coatings containing 25 phr ATBN suggest unusually high Tg values and appear anomalous. The ultimate Tg (72°C) of the coating is not realized until a TBA scan at 180°C is made (not included in Figure 1).

Adhesion

Adhesion can be quantitatively measured with an instrument known as the ELCOMETER Adhesion Tester. It is designed to measure the force, in a tensile failure mode, required to remove a coating from its substrate. To reinforce its authenticity, the ELCOMETER (Scale range 0-1000 psi) was calibrated against an INSTRON Tester. There is close agreement between the two instruments at intermediate and higher adhesion values. See Figure 2.

Elastomer-modified epoxy coatings have demonstrated better adhesion to sandblasted cold-rolled steel than non-elastomer containing analogs. Table X summarizes these results for both clear and pigmented coatings with either CTBN or ATBN as elastomeric modifier. Recent work (58) suggests that these adhesion levels may be improved further with organo-functional silane coupling agents.

Impact Strength

Impact resistance of epoxy coatings, as determined by a reverse Gardner impact test, is enhanced by liquid polymer addition independent of functional type. Again, these results are consistent with both unpigmented and pigmented model coatings. Although

TABLE IX

IMPACT STRENGTH OF ATBN MODIFIED EPOXY COATINGS
(CURE - 7 DAYS AT R.T.)

	1	2	3	4
DGEBA (EEW, 525)	100	100	100	100
Methylisobutyl ketone	32	33	34	34
Leveling Agent	1.7	1.7	1.7	1.7
Polyamide (Amine Value, 238)	50	48.5	47.1	45.6
Cellosolve	16	17	18	18
Xylene	33	34	35	36
ATBN (1300X16)	-	5	10	15
Reverse impact, in-lbs; 5/8" dart				
Room temperature	230	290	290	320
4°F (-16°C)	175	205	205	290

TABLE X

ADHESION OF RUBBER MODIFIED EPOXY COATINGS TO
SAND BLASTED COLD ROLLED STEEL
(CURE - 7 DAYS AT R.T.)

	1	2	3	4	5	6
DGEBA (EEW, 525)	100	100	100	100	100	40.1
Methylisobutyl ketone	33	33	63	64	65	115
Xylene	7	7	-	-	-	-
Cellosolve	13	13	-	-	-	-
Leveling Agent	5	5	5	5	5	5
Titanium dioxide	-	-	150	157	150	148
CTBN-Epoxy Adduct[1]	-	-	-	-	-	69.9
DETA - DGEBA Adduct (Amine Value, 280)	50	46.7	-	-	-	-
Polyamide (Amine Value, 238)	-	-	50	47.1	50	37.8
Cellosolve	-	-	32	33	32	59
Xylene	-	-	69	70	67	123
ATBN (1300X16)	-	15	-	10	-	-
ELCOMETER Adhesion, psi	820	>1000	730	>1000	730	>1000

[1]14.3% CTBN-1300X13; EEW, 1035

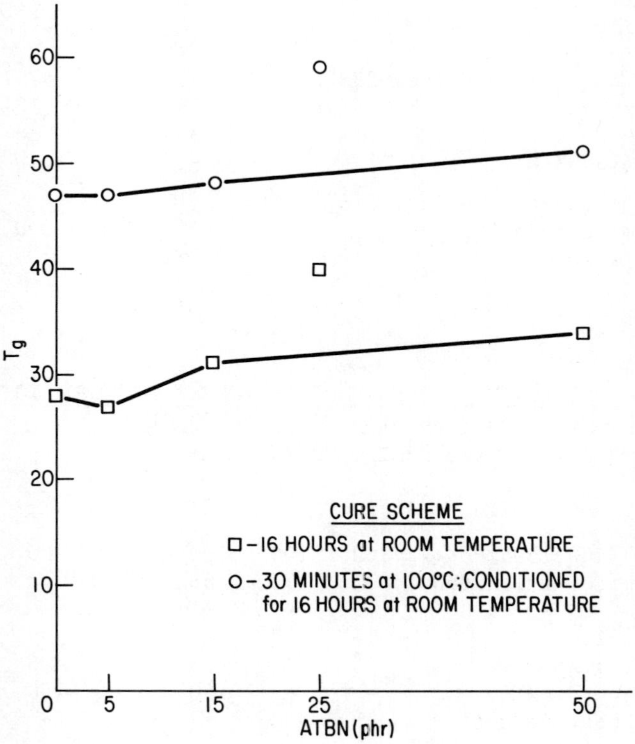

Figure 1. Resin glass transition temperature vs. ATBN concentration.

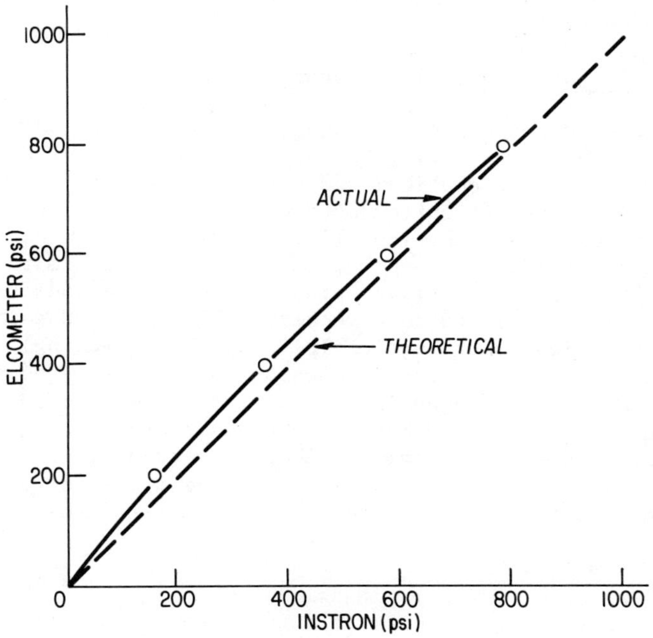

Figure 2. Calibration of Elcometer adhesion tester with Instron tester.

ATBN-modified epoxy/polyamide coatings (Table IX) exhibit excellent impact strength, CTBN-modified versions merit special attention (Table XI).

As previously mentioned in this chapter, carboxyl-reactive nitrile liquids appear to be the preferred modifiers for latent epoxy structural film adhesives. One of the major components used in formulating these adhesives is a solid epoxy resin of similar molecular weight (1000). This limited coatings study suggests that similar elastomer-modified epoxy resins are equally impressive in solution epoxy coatings. In addition, Table XI/Recipe 3 features a high impact coating modified with both liquid and solid carboxylated nitrile elastomers.

Salt Spray and Chemical Resistance

In comparing two coatings having the same generic classification, the one having uniformly better adhesion to metal may very likely provide greater chemical and corrosion resistance. This accepted belief, not without exception however, is illustrated in solventless epoxy model coatings (59) in which the unmodified coating shows considerable spot rusting (1-2 ASTM D-610 rating) in less than 200 hours in the salt spray chamber. The modified coating, previously determined to have higher ELCOMETER adhesion values, has only slight spot rusting (8-9 ASTM D-610 rating). Supporting the claim that improved adhesion can partially equate to improved chemical resistance is the caustic resistance information found in Table XII. In another solventless epoxy coatings model (60) utilizing an aromatic amine adduct as hardener, liquid nitrile rubber inclusion has been found to provide generous forward and reverse impact strength improvements in three-mil films.

Summary

The chemical literature reveals relatively new and varied interests for nitrile elastomer-modification of epoxy resins in diverse areas of coatings and primers. Desirable properties such as impact resistance, mandrel bend and adhesion improvement are attainable with little or no sacrifice in critical film properties. In some instances, it is documented that proper elastomer modification of select epoxy coatings will enhance corrosion and moisture resistance. This may relate to better film/substrate adhesion durability. The combined literature, journal and patent, has led to continuing study of nitrile elastomer modified epoxy coatings.

TABLE XI

IMPACT STRENGTH OF CARBOXYLATED NITRILE MODIFIED EPOXY COATINGS (CURE - 7 DAYS AT R.T.)

	1	2	3
DGEBA (EEW, 525)	100	40.1	-
CTBN - Epoxy Adduct[1]	-	69.9	-
Carboxylated Solid Nitrile - Epoxy Adduct[2]	-	-	166.7
Titanium dioxide	150	148	-
Methylisobutyl ketone	65	115	92.8
Leveling Agent	5	5	5
Polyamide (Amine Value, 238)	50	37.8	38.1
Cellosolve	32	59	47.6
Xylene	67	123	96.7
Reverse impact, in-lbs 5/8" dart			
Room temperature	79	>320	>320
Pencil hardness	F	HB	B

[1] 14.3% CTBN-1300X13; EEW, 1035

[2] 10% CTBN-1300X13; 6% 1472; EEW, 460; 72% Solids

TABLE XII

CHEMICAL RESISTANCE OF CTBN MODIFIED EPOXY COATINGS (CURE - 7 DAYS AT R.T.)

	1	2
DGEBA (EEW, 525)	100	40.1
CTBN - Epoxy Adduct[1]	-	69.9
Titanium dioxide	150	148
Methylisobutyl ketone	65	115
Leveling Agent	5	5
Polyamide (Amine Value, 230-246)	50	37.8
Cellosolve	32	59
Xylene	67	123
Chemical Resistance		
Glacial acetic acid, 30 minutes	Fail	Fail
10% Hydrochloric acid, 1 week	Fail	Fail
5% Aqueous caustic, 2 weeks	Fail	Pass
Boiling water, 1 hour	Pass	Pass
MEK, 4 hours	Pass	Pass
Toluene, 1 week	Pass	Pass

[1] 14.3% CTBN-1300X13; EEW, 1035

Literature Cited

1. Hoffman, W. "Nitrile Rubber (Review)" RC & T, Technical Review Supplement 1964, 37 (2-2).
2. Brown, H. P. RC & T 1963, 36 (4), 931.
3. Drake, R. S.; McCarthy, W. J. Rubber World 1968 (Oct.).
4. Riew, C. K. RC & T 1981, 54 (2).
5. BF Goodrich Chemical Group, "Toughen Epoxy Resins with Hycar Reactive Liquid Polymers" 1980, RLP-2.
6. Shelley, R. R.; Clarke, J. A. British Patent 1 461 127, 1977.
7. Pavlíková, A.; Luňák, S.; Lednický, F.; Dušek, K. Chemický Průmysl 1980, 30/55 (11), 591-597.
8. Lee, B. L.; Lizak, C. M.; Riew, C. K. 12th National SAMPE Tech. Conf. 1980, 12, 1116-1126.
9. Shelley, R. R.; Clarke, J. A. British Patent 1 461 127, 1977.
10. Riew, C. K. RC & T 1981, 54 (2).
11. BF Goodrich Chemical Group, "Improving Epoxy Resins" 1981, RLP-3.
12. Riew, C. K.; Rowe, E. H.; Siebert, A. R. "Toughness and Brittleness of Plastics", edited by D. R. Deanin and A. M. Crugnola, Advances in Chemistry Series, 154, American Chemical Society, 326 1976.
13. Hunston, D. L.; Bitner, J. L.; Rushford, J. L.; Rose, W. S.; Riew, C. K. "Adhesion and Adhesives: Science, Technology and Applications", The Plastics and Rubber Institute (London) 1980, Grey College Durham, Paper 14.1.
14. Manzione, L. T.; Gillham, J. K.; McPherson, C. A. J. Appl. Polym. Sci. 1981, 26, 889-919.
15. Daly, J.; Pethrick, R. A.; Fuller, P.; Cunliffe, A. V.; Datta, P. K. Polymer 1981, 22, 32-42.
16. Sayre, J. A.; Assink, R. A.; Lagasse, R. R.; Ibid. 87-94
17. Sohn, J. E. "Morphology of Solid Uncured Rubber-Modified Epoxy Resins", 181st National Meeting, ACS, ORPL, 44 (March, 1981).
18. Drake, R. S.; Siebert, A. R. SAMPE Q. 1975, 6 (1).
19. Bascom, W. D.; Bitner, J. L.; Moulton, R. J.; Siebert, A. R. Composites 1980 (1), 9-18
20. Moulton, R. J. "Advanced Technology in Materials Engineering", SAMPE International Conference 1981, Cannes, France.
21. Klapprott, D. K.; Paradise, D. L. US Patents 3 678 130 and 3 678 131, 1972.
22. Gross, M. E.; Weber, C. "Materials Review-'75" SAMPE Conference, 7 1975 (Oct.).
23. Flickinger, J. H. J. of the Adhesives and Sealants Council 1974, III (1), Park Ridge, Ill.
24. Lees, W. A. "Adhesion and Adhesives: Science, Technology and Applications", The Plastics and Rubber Institute (London) 1980, Grey College Durham, Paper 17.1.
25. Bolger, J. C. Adhesives Age, 1980 (Dec.).

26. Paul, N. C.; Richards, D. H.; Thompson, D. Polymer 1977, 18 945-950
27. Bascom, W. D.; Hunston, D. L. "Toughening of Plastics", The Plastics and Rubber Institute (London) 1978, Paper 22.1.
28. Kinloch, A. J.; Shaw, S. J. J. Adhesion 1981, 12, 59-77.
29. Walker, J. W.; Richardson, W. E.; Smith, C. E. Mod. Plast. 1976 (May).
30. Furney, M. F. "Materials Review-'75" SAMPE Conference, 7 1975 (Oct.).
31. Creed, K. E. "Epoxy Resin-CTBN Encapsulant", Contract AT-(29-2)-656 USA EC, May 14, 1974.
32. Hussain, A.; McGarry, F. J. "Toughening of Anhydride Cured Epoxy Resins" Res. Report R80-2, 1980 M.I.T., Cambridge, MA.
33. Hill, J. W. SPE ANTEC Preprints, S. F., CA. 1974 (May).
34. Navjar, D. J. US Patent 3 892 819, 1975.
35. Waters, W. D. US Patent 3 928 491, 1975.
36. Currieo, R. A. SPI-30th RP/Composites Conference, 1975, Section 13-B.
37. Hawthorne, K.; Stavinoha, R.; Craigie, L. 32nd SPI RP/Composites Institute, 1977, Section 5-E.
38. Siebert, A. R.; Riew, C. K., "The Chemistry of Rubber-Toughened Epoxy Resins", 161st National Meeting, ACS, ORPL, March, 1971.
39. McPherson, C. A. US Patent 4 121 015, 1978.
40. Ciba Geigy (Italia) SpA, "Product XG-75", Plastics Materials Lab Bulletin.
41. Spencer Kellogg, Inc. - Technical Data Sheets TD-79104, TD-7866 (1980).
42. Lewis, R. B. US Patent 4 176 142, 1979.
43. McKown, A. G. US Patent 3 655 818, 1972; Canadian Patent 935 246, 1973.
44. Nitto Denko,Japanese Patent 56-122823, 1980.
45. Ariga, N. US Patent 4 085 018, 1978.
46. Diener, S. L. 'Development of Improved Electrodeposited Inhibiting Primers', AFML-TR-79-4073, June, 1979 Final Report.
47. Tsuchiya, Y.; Ichikawa, A.; Yamasoe, K. US Patent 4 253 930, 1981.
48. Tomegawa Paper Mfg KK, Japanese Patent 55-125166, 1980.
49. Athey, R. D. J. Coatings Tech. 1982, 54 (690), 47-50.
50. Mendelsohn, M. A. US Patent 4 298 656, 1981.
51. Hoechst Work Hamburg, Data Sheet M2097 for Beckopox EH 629 (Germany).
52. Numata, S.; Kinjo, N. Polymer J. 1982, 14 (8), 671-673.
53. Tilley, C. G.; Sinclair, J. H. from "Liquid Polymers Symposium", University of Surrey, Sept. 26/27, 1972.
54. Kingsley, G. ed. "Market Survey and Users Reference", Graduate School of Business Administration, Harvard University, Cambridge, MA. July, 1959.
55. Lee, H.; Neville, K. "Handbook of Epoxy Resins", McGraw-Hill, New York, 1967, pp. 7/15-26.

56. Bruins, P. F., ed. "Epoxy Resin Technology" Interscience, 1968.
57. Manzione, L. T.; Gillham, J. K.; McPherson, C. A. J. Appl. Polym. Sci. 1981, 26, 889-905.
58. Walker, P. J. Oil Co. Chem. Assoc. 1982, 65, 415-423.
59. Spencer Kellogg, Inc. - Technical Data Sheet TD-7866 (1980).
60. Young, R. G.; Howell, W. R. Mod. Paint and Coatgs., Mar., 1975.

RECEIVED March 2, 1983

2

Elastomeric Polysiloxane Modifiers for Epoxy Networks

Synthesis of Functional Oligomers and Network Formation Studies

J. S. RIFFLE—Union Carbide Corporation, S. Charleston, WV 25303
I. YILGOR, C. TRAN, G. L. WILKES, and J. E. McGRATH [1]—
Virginia Polytechnic Institute and State University, Department of Chemistry, Chemical Engineering, and Polymer Materials and Interfaces Laboratory, Blacksburg, VA 24061
A. K. BANTHIA—Indian Institute of Technology, Materials Science Centre, Kharagpur 721302, India

Novel elastomeric polysiloxane modifiers for epoxy networks have been synthesized and characterized. In addition, curing studies of conventional epoxy resins incorporating these oligomers have been conducted. Structures were prepared having either epoxide, primary amine and/or secondary amine endgroups. The polymerization step for the siloxanes consisted of a base catalyzed equilibration of the appropriately functionalized disiloxane and octamethylcyclotetrasiloxane. In the case of hydroxy piperazine terminated modifiers, polymers were first synthesized with epoxy endgroups. These endgroups were then subsequently capped with an excess of piperazine prior to curing. The oligomers were characterized by FTIR, 'H and ^{13}C NMR, GPC, endgroup analysis and vapor pressure osmometry. The secondary amine terminated oligomers basically act as linear modifiers. By contrast, the aminopropyl functional siloxanes produce a crosslinked network. Piperazine terminated oligomers show much better compatibility with the epoxy resins compared to the aminopropyl terminated oligomers. This is especially true if the modifier contains an amide group. The curing reactions of the epoxy resins (EPON resin-828) using these oligomers and a cycloaliphatic diamine (PACM-20) were followed under a variety of conditions by DSC. Percent conversion versus time plots showed that reactivities of the functional siloxanes are higher than the cycloaliphatic amine at comparable temperatures. X-ray photoelectron spectroscopy shows the surfaces of the polysiloxane modified materials to be predominantly siloxane even at quite low bulk levels of the reactive oligomers. Preliminary mechanical property studies are encouraging and detailed morphology-property studies are continuing.

Epoxy resins are one of the most important classes of thermosetting polymers. They are widely used for many important

[1] To whom correspondence should be directed.

0097–6156/83/0221–0021$09.50/0

applications such as coatings, adhesives, reinforced plastics and matrix resins for advanced composite materials (1-4). If the network structure is properly generated, epoxy materials can exhibit excellent corrosion and solvent resistance, good adhesion to many substrates and adequate electrical properties. However, in contrast to such desirable characteristics, epoxy networks are typically rather brittle and display rather low fracture toughness. Moreover, the nature of the crosslinking agent and its concentration may also induce relatively high water absorption characteristics. For example, a structural epoxy system may be cured with aromatic amines. The resulting nitrogen containing crosslink provides sites for hydrogen bonding to water. Absorption of water has the effect of plasticizing these materials and thus deteriorating one of their more desirable characteristics. Moreover, the lack of toughness has limited the application characteristics of these systems. Many studies have been conducted which have attempted to improve the fracture and/or impact energies of the networks, while at the same time retaining the other properties. In order to do this, one must closely define the parameters such as the modifier structure, molecular weight, solubility parameter, and concentration of the elastomer. In addition, the structure and reactivity of the hardener and the curing agent must be carefully controlled (5-8). It is important initially to define elastomer systems that are sufficiently "miscible" with the uncured epoxy formulation that they can be adequately dispersed. During the formation of the network (e.g. the curing step), if the solubility parameter of the elastomer and epoxy matrix structure differ sufficiently, it has been possible to develop phase separated elastomer epoxy materials. Thus, small domains of the rubber phase which would be the minority component may be dispersed in the continuous matrix composed of the epoxy resin and the hardener. This dynamic formation of the multiphase structure is very sensitive to reaction conditions and the characteristics of the rubber, such as its molecular weight. Therefore, the functionality and reactivity of the elastomer as compared with the hardener are also important parameters.

Previously, most of the research on the toughening of epoxy networks has utilized the commercially available carboxy terminated butadiene-acrylonitrile (NBR) rubbers (6, 9-11). The carboxy materials are capable of interacting with the epoxy groups to generate a hydroxyester as shown in Scheme 1. Thus, this route has provided a mechanism for chemically bonding the elastomeric component to the rigid matrix. The chemically bonded system then presumably should provide a relatively stable linkage between the modifier and the matrix. If the rubber particles are of appropriate size, they may modify the properties of the epoxy system in a somewhat analogous way to the toughening of thermoplastics (23). It is thought, for example, that the mechanism of craze yielding may be influenced by the presence of the elastomer

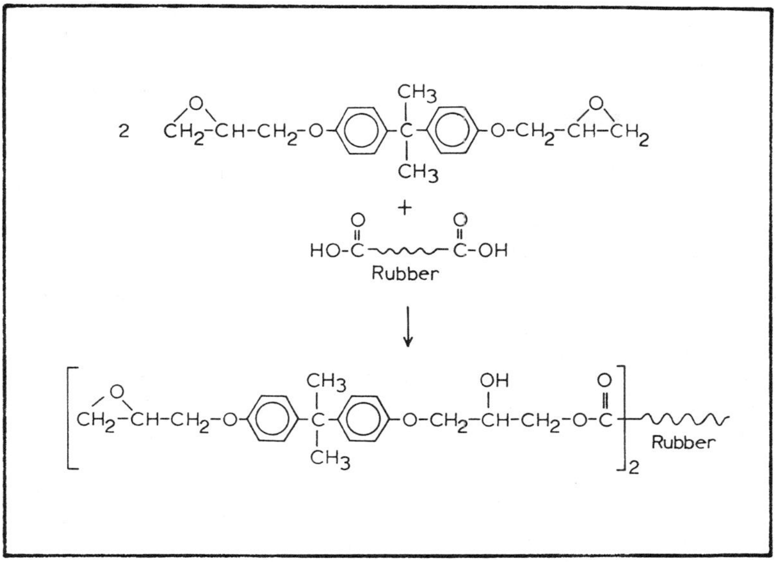

Scheme 1

particles. More recently, the use of butadiene-acrylonitrile rubbers terminated with amino groups has been investigated. These types of modifiers have been reviewed recently by Riew (10) and have also been discussed in detail by Drake et al. (11).

The butadiene acrylonitrile rubbers have clearly been successful to a considerable extent in improving the toughness characteristics of epoxy networks. However, since the butadiene component of the elastomer contains unsaturation, it would appear to be a site for premature thermal and/or oxidative instability. One would imagine that excessive crosslinking could take place with time which would detract from the otherwise desirable improvements accomplished with these structures.

We have been interested for some time in the chemistry and structure of polysiloxane containing systems. We suggest that some of the important characteristics of siloxane structures, such as their thermal and oxidative stability, low glass transition temperature, hydrophobic character and low surface energies could perhaps render them useful as epoxy modifiers. In order to do so, however, one would have to consider the questions of functionality, both with respect to type and concentration and also the miscibility or solubility of such hydrophobic nonpolar materials in the typically aromatic based epoxy precursors. Thus, the definition of such functional oligomers and their network forming characteristics appear to be a logical starting place for these investigations. This paper will describe our initial studies which have been concerned both with the synthesis of appropriately functionalized siloxane oligomers, their chemical characterization and with the development of the siloxane modified networks. Future papers will discuss the property modifications and characteristics in more depth.

EXPERIMENTAL

Synthesis of the Epoxy Terminated Disiloxane Precursor

The epoxy disiloxane was made in our laboratory by the hydrosilation reaction of allylglycidylether (Aldrich) with dimethylchlorosilane which was obtained from Petrarch. The reaction was conducted as a 50% solution in toluene, and was catalyzed by chloroplatinic acid at a concentration of approximately 10^{-3} moles per mole allylglycidylether. The reaction proceeded rapidly at 35°C as evidenced by a pronounced exotherm. Although it is felt that most of the reaction is completed very quickly, perhaps within five minutes or less, heating (35°C) was continued for approximately one hour. The molar ratio of the allylglycidylether to the dimethylchlorosilane was about 1.3 to 1. The excess allylglycidylether was utilized to insure that all of the Si-H bonds were in fact reacted. The product was purified by vacuum distillation at 1.0 torr and 90°C. The chlorine substituted silane was then hydrolyzed in water to produce the desired disiloxane. The resulting product was colorless, transparent and was fully characterized by proton NMR as being consistent with the desired structure. Moreover, the boiling point was shown to match the value given in the literature (12).

Preparation of the Siloxanolate Equilibration Catalyst

In order to induce the polymerization of the cyclic tetramer in the presence of these functional siloxanes, a siloxanolate catalyst was prepared from the reaction of 4 moles of the cyclic tetramer (D_4) with 1 mole of pure tetramethylammonium hydroxide. In one approach, the solid was dispersed in the liquid D_4 and the reaction was conducted under an argon stream for about 48 hours at 60-70°C. The rapid argon flow was sufficient to dehydrate the system. The reaction when conducted under these conditions is somewhat complex. After about 5 hours we have observed that the system appears rather opaque and that the viscosity is increased significantly. As a function of time, the viscosity begins to decrease and the material, while not becoming perfectly transparent, does increase in translucency relative to the original opaque mixture. The catalyst produced in this manner is sufficiently active for the synthesis of all the oligomers discussed in this paper. However, for other oligomers, such as silylamines, additional refinements appear to be necessary and will be discussed in detail in future papers. Briefly, utilization of an azeotroping solvent such as hexane is desirable. Higher D-4/base ratios may also be used.

Equilibration Polymerization of Epoxy Functional Disiloxanes

The epoxy terminated siloxanes were produced via the so-called equilibration polymerization of the previously mentioned disiloxane discussed in Section 1 above through the use of the quaternary ammonium siloxanolate catalyst discussed in Section 2. The molecular weights of the siloxanes were varied via controlled ratios of the cyclic tetramer (D_4) to the epoxy disiloxane "end-blocker". The reactions were conducted at 80°C for various periods of time that ranged to as much as 44 hours. The catalyst concentration of the siloxanolate discussed in 2 above was varied from approximately 0.1% to as high as 1% by weight based on the total concentration of reactants. The reaction was followed by both GPC and also titration methods which are described more fully in a later section of the experimental. After the appropriate reaction time, the temperature was increased to about 150°C for approximately another 3 hours. This time is believed sufficient to decompose the so-called transient catalyst (13) and thus inactivate the equilibration process. At this point, the oligomer also contains cyclics. In order to remove the cyclics, the crude oligomers were vacuum stripped under mechanical pump vacuum at approximately 0.5 to 1 torr and about 100°C. The stripped materials are then characterized further as discussed below. Typically, the stripped materials are perfectly transparent. An alternate method for purification of the crude oligomers is to extract the equilibrium cyclics with a solvent such as methanol which will dissolve the cyclic but will not dissolve oligomers of ~1000 molecular weight or higher.

Equilibration of Aminopropyl Terminated Disiloxanes

The aminopropyl terminated disiloxane was obtained from Petrarch, although one could in principle prepare this as discussed earlier by hydrosilation procedures. The equilibrations were conducted by reacting the cyclic tetramer in the presence of bis(aminopropyl)tetramethyldisiloxane and about 0.5 weight percent of the siloxanolate catalyst at 80° for 44 hours. These conditions were selected on the basis of earlier extensive studies with the epoxy terminated oligomer. Again, the catalyst was decomposed by reaction at 150°C for 3 hours prior to the extraction of the cyclics. The resulting amino terminated oligomer molecular weight is governed largely by the ratio of the cyclic tetramer to the disiloxane. It was worked up in an analogous manner to the epoxy systems, e.g. either by vacuum stripping or by treatment with methanol.

Equilibration of Carboxypropyl Terminated Siloxanes

The carboxypropyl functional disiloxane was obtained from Silar Laboratories, Scotia, New York. Unlike the previous two equilibration systems, it was necessary here to utilize an acid catalyst. We investigated several systems, but here we described how we have utilized trifluoroacetic acid. The carboxy terminated disiloxane and the D_4 were mixed in appropriate ratios and the temperature was brought to 60°C. At this point about 14% of pure trifluoroacetic acid was introduced. The reaction was allowed to proceed for 24 hours and one could observe the viscosity increasing slowly as a function of time. Other time intervals were investigated and 24 hours appeared to be approximately the optimum period for this particular synthesis. The reaction was also followed by a GPC method as discussed below. In this case, the product was purified by repeated washing of the crude oligomer with distilled water in order to remove the excess acid catalyst. The washings were continued until one could no longer observe any fluorine present, especially as judged by ^{19}F NMR experiments. After the oligomer has been sufficiently washed with distilled water, it was extracted from the aqueous suspension via diethyl ether. The diethyl ether was removed by a rotary vacuum stripping apparatus. The residual cyclic oligomers were devolatilized as indicated for the systems above, that is at about 0.5 torr and approximately 100°C.

Synthesis of Piperazine Terminated Disiloxanes from the Reaction of Carboxy Functional Disiloxane and Aminoethyl Piperazine (AEP)

The carboxy terminated disiloxane discussed above was mixed with an excess aminoethyl piperazine (Aldrich) and subsequently heated to 160°C for 24 hours. The reaction was followed by studying the FTIR spectrum which indicated the formation of the amide bond as a function of time and the disappearance of the carboxyl group. In addition, the material was characterized via titration methods. The titrations included both those for the carboxyl group and the secondary piperazine endgroups. Before the titrations were conducted, the excess amine was stripped from the crude reaction material under high vacuum. The titration of the stripped reaction product indicated the expected molecular weight. This was further confirmed by molecular weight measurements by vapor pressure osmometry in chloroform solvent at 37°C. Both the FTIR spectra (10) and the absence of a second endpoint in the potentiometric titrations indicate that only the secondary piperazine amine endgroup is present.

Equilibration of Piperazine Capped Disiloxanes

The piperazine capped disiloxane as discussed in 6 above could easily be equilibrated under exactly identical conditions to those discussed for the primary amine system earlier, that is, the disiloxane, cyclic tetramer, and 0.5 weight percent of the siloxanolate catalyst were heated to 80°C for 44 hours. The work-up procedure was identical, that is, the catalyst was deactivated at 150°C for 3 hours. The excess cyclics were then stripped under vacuum (0.5 torr, 100°C) and the material was further characterized by amine endgroup titration.

Piperazine Capping of Epoxy Terminated Siloxane Oligomers

An additional way to produce a secondary amine group utilized the reaction of 8 moles of piperazine with 1 mole of the epoxy terminated siloxane oligomer. This reaction was conducted in dioxane at 60° for about 24 hours. Excess piperazine was removed by washing the oligomer extensively with distilled water. Since the piperazine is water soluble and the oligomer is not, the final system was easily purified to very low levels of residual piperazine in this manner. The product was characterized by titration of the amine endgroups.

Synthesis of the Modified Epoxy Networks

For the purposes of this study, we have utilized commercially available epoxy oligomers based on bisphenol-A and epichlorohydrin. These particular oligomers were kindly supplied by the Shell Development Company (EPON resin-828). The basic route to thermoset formation that we have utilized has been to react stoichiometric quantities of diamines with the epoxy network precursors. The amine that we have primarily worked with is bis(para-aminocyclohexyl)methane which was generously supplied by the DuPont Company under the trade name PACM-20. Test specimens were usually prepared in silicone molds which were purchased from Dow Corning under the trade designation of #3110 Silicone Rubber. The conditions for network formation were investigated for the control systems and it was decided to utilize two different conditions, a so-called one-step and a two-step process. In the one-step process, the two components were first degassed separately in a vacuum oven at about 30 torr and 50°C. Once they were sufficiently degassed they were quickly mixed to form a homogeneous solution and cast into the silicone molds at 160°C. The reaction was allowed to continue under these conditions for two hours, which appears to be sufficient time to basically complete the reaction. It is also above the glass temperature of the final network structure. In the two-step process, the epoxy (EPON resin-828) and the modifier were first mixed at 60°C for about 2 hours. Subsequently, the degassed PACM-20 was added and the mixture was then poured into the silicone mold at 160°C for 2 hours as was done for the one-step process. Note that the two-step process has been exclusively utilized for the piperazine capped epoxy siloxane modifiers. In both processes, the materials were removed from the mold after the 2 hour curing cycle and cooled down to room temperature. The networks prepared in the manner described above were utilized for thermal and mechanical studies but not for ESCA spectroscopic studies. Those samples were prepared differently as discussed below.

Characterization

Titrations

The functional epoxy oligomers and their precursors were characterized by a number of methods. The titration method utilized a Fisher Model II automatic titrator which was capable of derivative measurements and utilized basically a potentiometric titration mode. The epoxy groups were titrated according to standard procedures discussed in the literature (14). Carboxy and amine groups were simply titrated with either base or acid which enabled one to observe appropriate potentiometric endpoints. Indicators were also utilized to complement the potentiometric studies. In the case of the epoxy oligomers, we used crystal violet; for the amine oligomers, bromophenol blue was employed. The titrations of the epoxy group via the potentiometric method did require a carefully purified solvent. We utilized chlorobenzene as a solvent which had been previously treated with concentrated sulfuric acid to remove any olefinic structures, followed by washing with water and careful fractional distillation.

NMR

Proton NMR was used routinely to confirm a number of the structures, especially those of the disiloxanes. In general, these measurements utilized a Varian EM-390 at room temperature with a methylene chloride or benzene chemical shift reference. The measurements were conducted in deuterated chloroform.

Gel Permeation Chromatography (GPC)

A Water's instrument model 440 was utilized at room temperature. The columns were microstyrogel of the following sizes (500Å, 10^3Å, 10^4Å, 10^5Å). Toluene solvent was used together with a differential refractive index (DRI) detector. The concentration of the oligomers in toluene was about 0.5 weight percent and the flow rate was approximately 1 ml/minute. The procedure described above appeared to be successful for the epoxy terminated oligomers and the carboxy terminated oligomers. Unfortunately, the aminopropyl and piperazine terminated oligomers showed anomalous behavior which we believe is a function of the absorption of the amine groups onto the packing materials. Therefore, our discussions will be limited to the epoxy and carboxy capped materials. The GPC did clearly distinguish between the disappearance of the cyclic tetramer and the formation of the oligomeric species during the course of the equilibration reactions discussed above.

Differential Scanning Calorimetry (DSC)

Thermal analysis measurements via DSC were utilized both to follow the curing reaction and to characterize the precursor and network materials. The general techniques for following reactions by differential scanning calorimetry are well established and can be reviewed in reference 15. We have also discussed the procedures for this type of an experiment in one of our earlier publications (16). A Perkin-Elmer Model 2 instrument was utilized.

X-Ray Photoelectron Spectroscopy (XPS, ESCA)

Surface analysis measurements were conducted on the siloxane modified epoxies using a DuPont 650 ESCA instrument in the MCA mode (Mg anode operating at approximately 22 milliamperes and 10kV). The samples for ESCA were prepared somewhat differently than those previously mentioned. In this case, the systems were mixed and cast directly onto the ESCA probe. This was done in an attempt to avoid the contamination of the surface with siloxane moieties that might be present in the silicone molds. Note that extreme care must be utilized in preparing such samples to avoid the somewhat ubiquitous contamination of siloxanes in the system. In fact, it was necessary to utilize a different laboratory to prepare the mixtures for these experiments. In this case, one could prepare control surfaces that were apparently free of extraneous siloxane contamination. The control samples were utilized together with various levels of siloxane modified systems, especially those derived from the piperazine capped epoxy oligomers. In addition, networks prepared with the epoxy oligomers themselves and the aminopropyl terminated oligomers were investigated via this surface technique.

ATR-FTIR Investigations

We have briefly investigated the cured and modified specimen via ATR-FTIR spectroscopy. A Nicolet MX-1 FTIR with a 45° KRS-5 crystal at an incidence angle of 50° was used. The FTIR spectra were obtained on several samples and were complementary to the ESCA in estimating the surface behavior of these modified systems.

RESULTS AND DISCUSSION

An idealized reaction of a diglycidyl ether of bisphenol A with a diamine is depicted in Scheme 2. Ideally one should

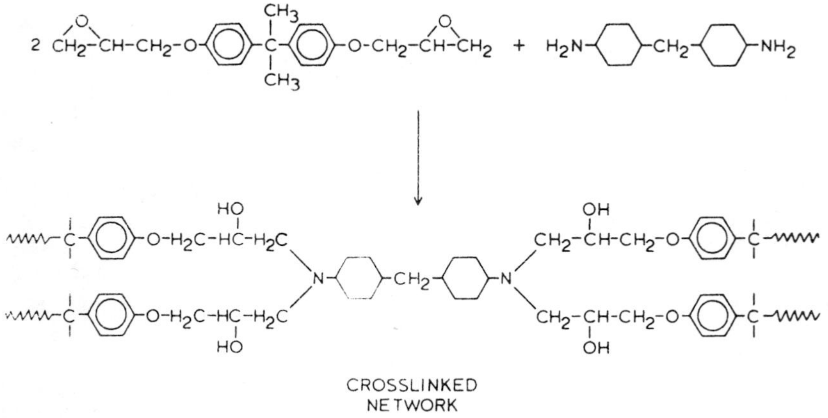

Scheme 2

generate four function points from each diamine molecule and therefore one could imagine a relatively uniform network density in such systems. While the question of network density is of great interest, it is not of primary concern to this paper, even though there may well be fewer than four functions per crosslink and conceivably the crosslinks may be somewhat non-uniform. We will discuss this point further in future papers that attempt to relate morphology to physical behavior. Our control systems were prepared from the EPON resin-828 and the cited bis(p-aminocyclohexyl)methane.

Synthesis of Functional Polydimethylsiloxane Oligomers

The oligomers utilized in this study to date have been prepared by the equilibration polymerization of a cyclic tetramer in the presence of functional disiloxanes utilizing a siloxanolate catalyst. The disiloxanes themselves were either synthesized in our laboratory or purchased. For example, in Scheme 3 we show one route that was utilized to prepare a diepoxy disiloxane endblocker. The reaction proceeds very rapidly at 35° as discussed in the experimental section. The reaction product can be further purified by vacuum distillation

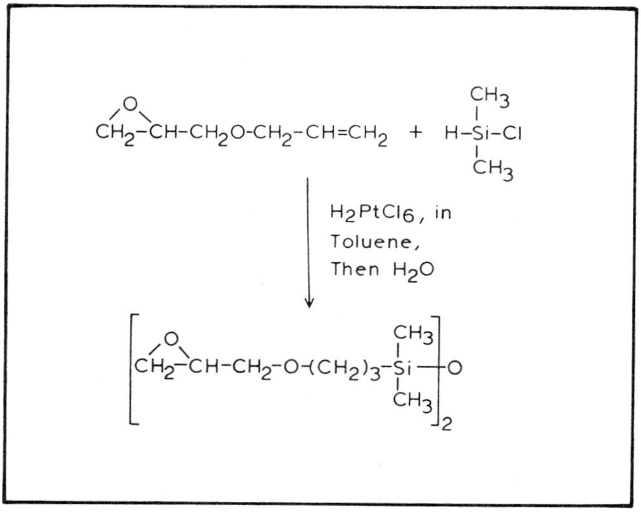

Scheme 3

at approximately 2 torr and 185°C. The product obtained displayed the correct proton NMR spectrum and is a relatively non-viscous clear liquid. The actual equilibration then to produce the epoxy terminated siloxanes is illustrated in Scheme 4. Here one introduces the disiloxane, the cyclic tetramer, and a catalytic quantity of the siloxanolate catalyst. The reaction temperature is 80°C for a period of time up to 44 hours. The reaction proceeds by attack of the siloxanolate on the silicon atom of the cyclic monomer to open the ring (initiation). The propagation step consists of further anionic attack on the cyclic monomers. As a function of time, the growing siloxanolate chain ends are able to attack the silicon oxygen bond in the disiloxane, thus becoming incorporated between the two epoxy endgroups. Since the silicon carbon bond in the disiloxane is not cleaved under these conditions, eventually at equilibrium the cyclic tetramer is incorporated to approximately 85% or higher into the chain which is capped perfectly with epoxy functionality. The residual D_4 is converted into cyclics which may be removed after catalyst deactivation by vacuum stripping as discussed in the experimental section. Alternatively, one may

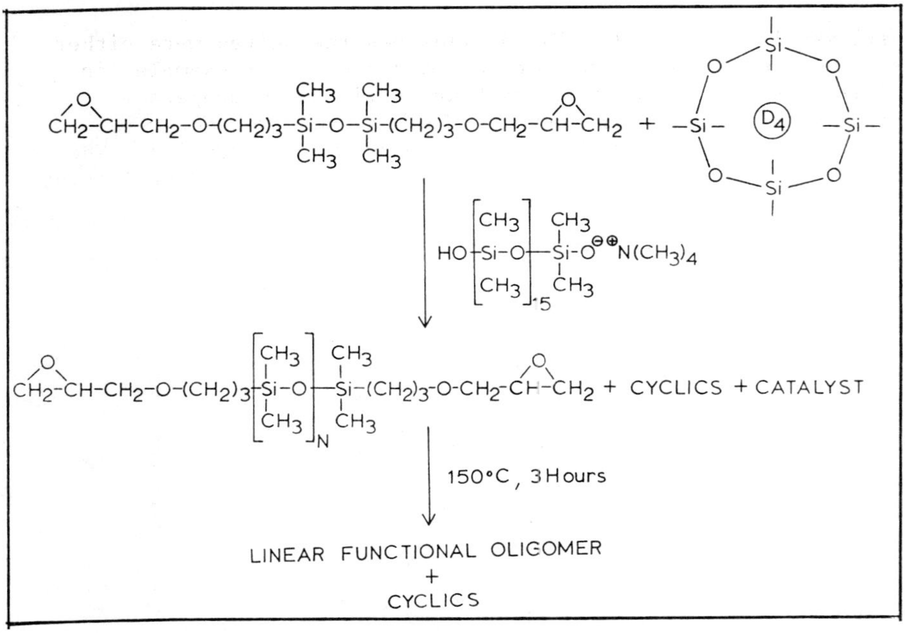

Scheme 4

wash out the cyclic oligomers with methanol. Note that higher alcohols may not be used since they will also dissolve the oligomer. Prior to extraction of the cyclics, the catalyst is deactivated by heating to about 150°C for 3 hours. The catalyst undergoes a Hoffman-like degradation to yield inert products, and hence it is termed a "transient catalyst" (13). It was very interesting and perhaps surprising to note that the epoxy terminated disiloxane could be effectively equilibrated with a basic siloxanolate-like catalyst. We were naturally concerned that the siloxanolate catalyst might also attack the epoxy groups on the disiloxane. Fortunately, this does not seem to be the case if one chooses the reaction conditions appropriately. For example, in Table I, we list some endgroup analyses that were conducted following the redistribution of the diepoxy system. We utilized, as discussed in the experimental section of this paper, a standard titration for the endgroups and compared the theoretical value on the basis of the ratio of the cyclic tetramer to the disiloxane to the actual equivalent weight which was measured by the titrations. This was also studied as a function of concentration of the siloxanolate catalyst ranging from about 0.1 to 1.0 weight percent. Various times up to 44 hours were investivated. As one may note, in most cases it appears that the equivalent weight is in good agreement with the theoretically predicted equivalent weight and any deviations are within experimental error. A control experiment was also

Table I

COMPARISON OF THEORETICAL AND EXPERIMENTAL EPOXY GROUP CONCENTRATIONS IN SILANOLATE EQUILIBRATED EPOXY FUNCTIONAL POLYDIMETHYLSILOXANE OLIGOMERS

$\langle M_N \rangle$ (THEORY)	EPOXY EQUIV. WT. (THEORY)	TITRATED EQUIV. WT.	WT. PERCENT EQUILIBRATION CATALYST	EQUILIBRATION TIME, (HOURS) AT 80°C
360, Dimer	180	190	--	--
Control	180	180	0.5%	22
5000	2500	2250	0.5%	22
	2500	2525	1.0%	22
5000	2500	2350	0.5%	44
	2500	2750	1.0%	44

conducted which used only the siloxane dimer and the silanolate catalyst (Table I, Control). After 22 hours at 80°C, there was no evidence that the epoxy group had been attacked by the silanolate. Therefore, on the basis of this result and some other data to be discussed later, we feel that the siloxanolate catalyst must be too weak a base to significantly attack the epoxide oxirane ring, especially when the reaction is conducted in the presence of the readily available siloxane bond. It is well known, for example, that alkoxides are much stronger bases than siloxanolates or silanolates and this seems to be a key point in this functional oligomer synthesis. We were also concerned with how one could follow equilibrations of this type. We chose to utilize gel permeation chromatography (GPC) data as a function of catalyst concentration as shown in Figure 1. We have shown here the elution time as a function of catalyst concentration after 44 hours of reaction time at 80°C. It appears that under the conditions we have utilized, 0.1% of the "bulk" prepared catalyst (e.g. no azeotroping solvent) is sufficiently active to initiate the equilibration. However, at 0.5% catalyst one sees clearly the disappearance of the cyclic tetramer and the formation of a most probable unimodal molecular weight distribution for the oligomer. Higher catalyst levels appears to produce somewhat higher rates. Complementary data of the effect of time on the epoxy terminated polydimethylsiloxane redistributions is shown in Figure 2. Again, one notes that certainly after 44 hours, the apparent equilibrium concentration of oligomers is reached.

The transient equilibration catalyst preparation is important. In order to synthesize the catalyst, one may react a quaternary ammonium hydroxide with the cyclic tetramer. An important variable to consider is the ratio of the cyclic tetramer to the quaternary ammonium salt. For this work, we have utilized a molar ratio of 4:1, that is, 4 moles of D_4 per mole of quaternary ammonium hydroxide. With the conditions utilized here, one presumes to have an average siloxane degree of polymerization of about 15 with a siloxanolate type end group. Alternatively, the silanol groups may couple and the siloxanolate may in fact be difunctional. The active catalyst chain end can be determined by a standard titration with for example, alcoholic HCl. It is important, of course, to dehydrate the system quite effectively for the purposes of this research. As discussed in the experimental section, we utilized initially merely a rather strong argon flow to achieve this dehydration. Other methods may also be used. For example, one may efficiently azeotrope the water with hexane or benzene. Note also that to equilibrate the disiloxanes discussed in this paper, even the catalysts dehydrated with argon work quite well. With more sensitive systems, such as silylamine functionalities, it is necessary to effectively dehydrate and to utilize a higher ratio of the cyclic tetramer to the base.

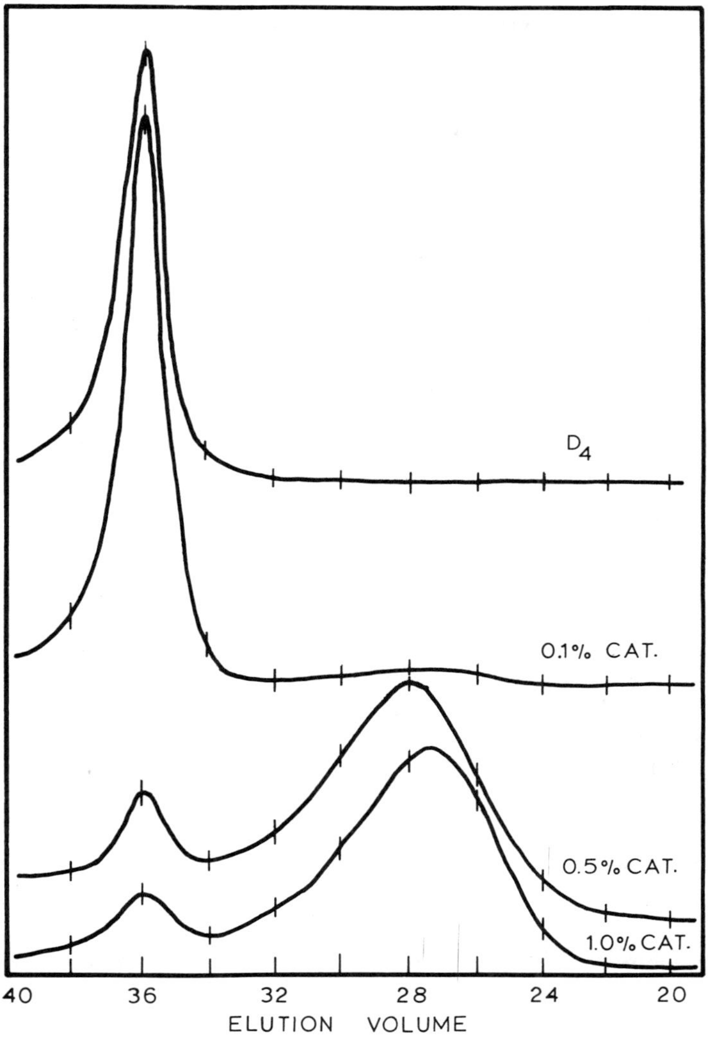

Figure 1. Effect of Catalyst Concentration on Epoxy Terminated Polydimethylsiloxane Equilibrations (44 hrs. reaction time).

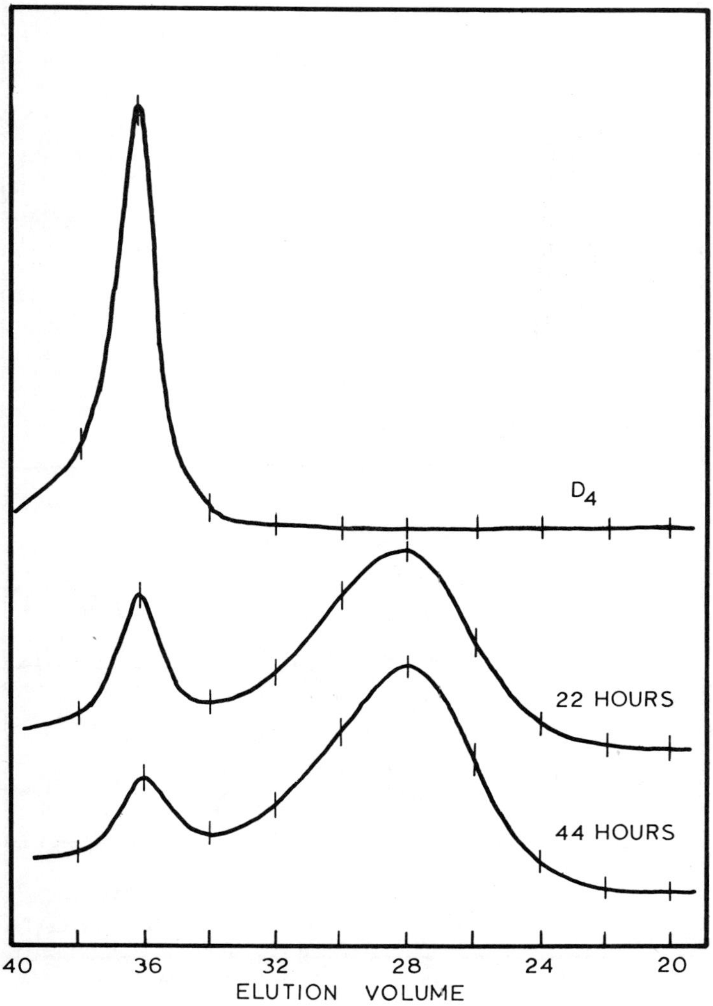

Figure 2. Effect of Reaction Time on Epoxy Terminated Polydimethylsiloxane Equilibrations (0.5% wt. Catalyst).

The equilibration of the aminopropyl terminated disiloxane was conducted in a very similar manner to the epoxy terminated disiloxane and is shown in Scheme 5. The reaction in this case could not be followed by our GPC unit since we noticed significant absorption of the oligomers onto the columns. Perhaps other columns would be more suitable and this question is being further explored. One approach that seems to be successful is to cap the end groups with trifluoroacetic anhydride, or benzoyl chloride.

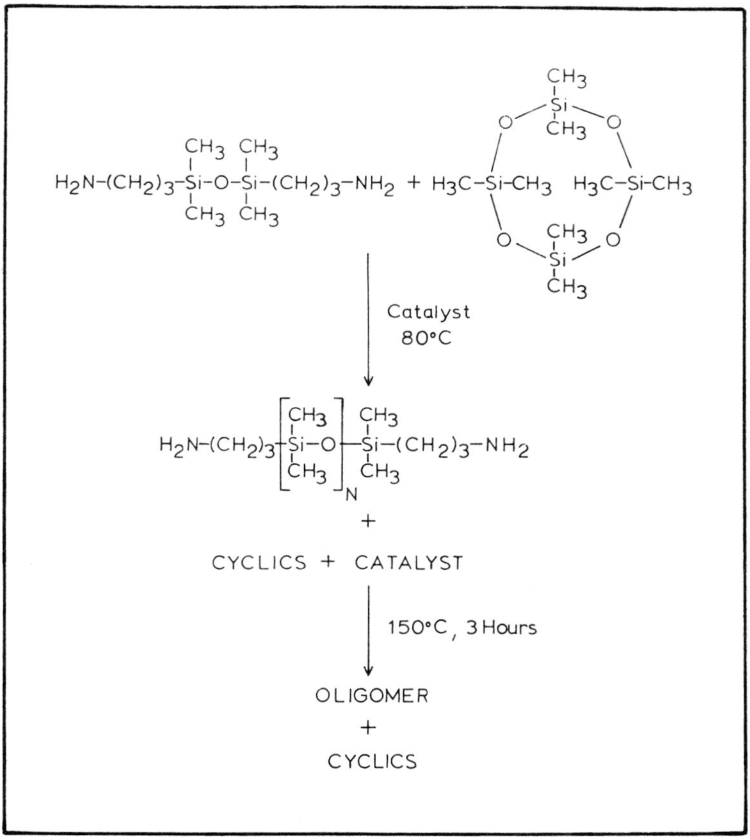

Scheme 5

A very similar equilibration was conducted as shown in Scheme 6 for the synthesis of secondary amine terminated siloxane oligomers. The amine terminated disiloxane here is a secondary

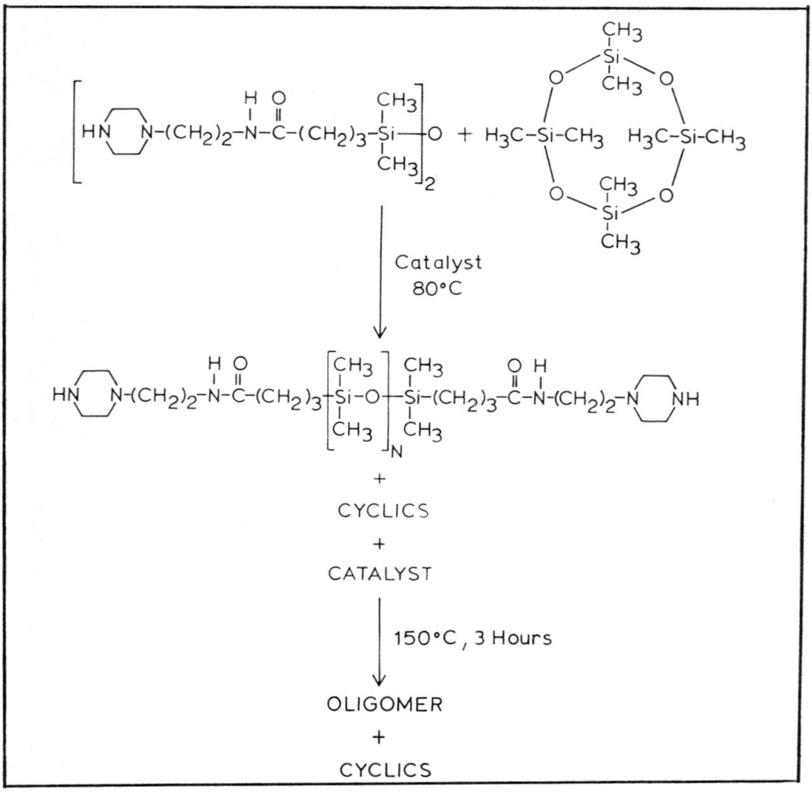

Scheme 6

amine which was prepared by the reaction of aminoethyl piperazine (AEP) with a biscarboxypropyldisiloxane. The carboxypropyldisiloxane was obtained from Silar Laboratories and presumably could be synthesized by hydrolysis of a nitrile derivative. We prepared the piperazine capped disiloxane by a bulk reaction at 160°C for 24 hours. The reaction product was purified by vacuum stripping of the aminoethyl piperazine. The resulting residual product gave only amine like titrations and a clearly defined amide band could be observed in the infrared. In addition, the infrared spectrum showed a negligible if any concentration of residual carboxyl groups. The synthesis and structure for the disiloxane is shown in Scheme 7. A different type of secondary amine functionality was prepared by capping of the epoxy terminated oligomer with an excess of piperazine as shown in Scheme 8. In this case, the epoxy oligomer was dissolved in dioxane and 8 moles of piperazine were added. The

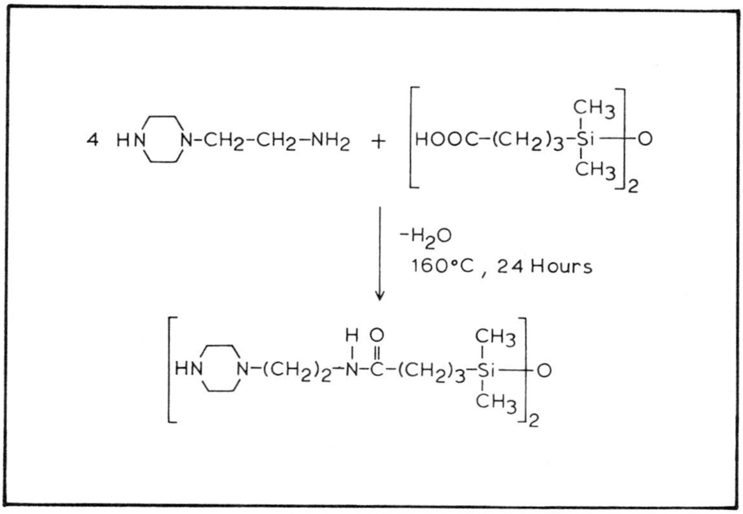

Scheme 7

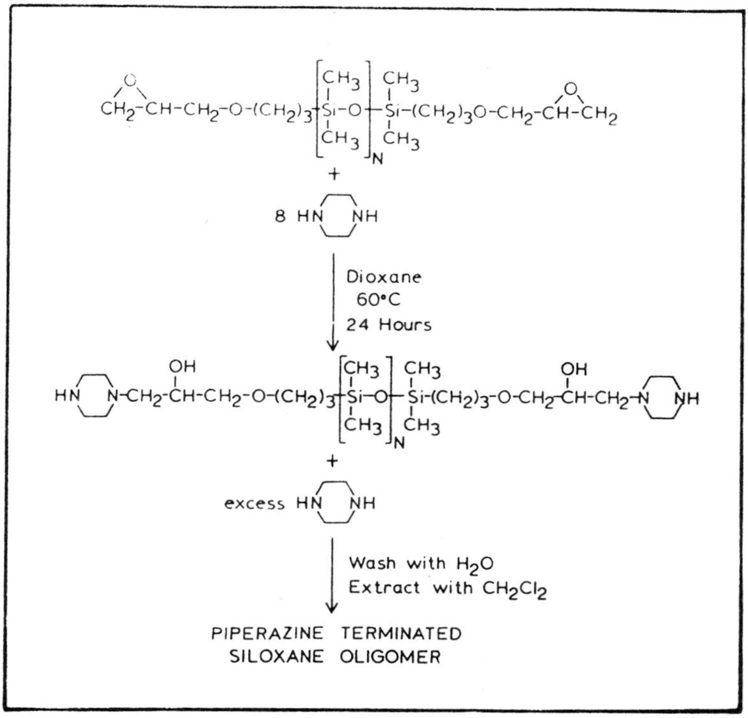

Scheme 8

material formed a homogeneous solution at 60°C and was allowed to react at that temperature for 24 hours. At this time, the reaction product was cooled down and thoroughly washed with distilled water. The water served to remove the excess piperazine. Next, the residual product was extracted with methylene chloride and vacuum stripped. The resulting derivative does in fact have a piperazine endgroup. However, unlike the previously discussed secondary amine this product also contains a hydroxyl group resulting from the cleavage of the epoxy group.

Network Formation Studies

The general structures of the siloxane modifiers that we have utilized in this study are shown for comparative purposes in Scheme 9. The overall curing reaction is depicted in Scheme 10. In this case we have illustrated the reaction by using the

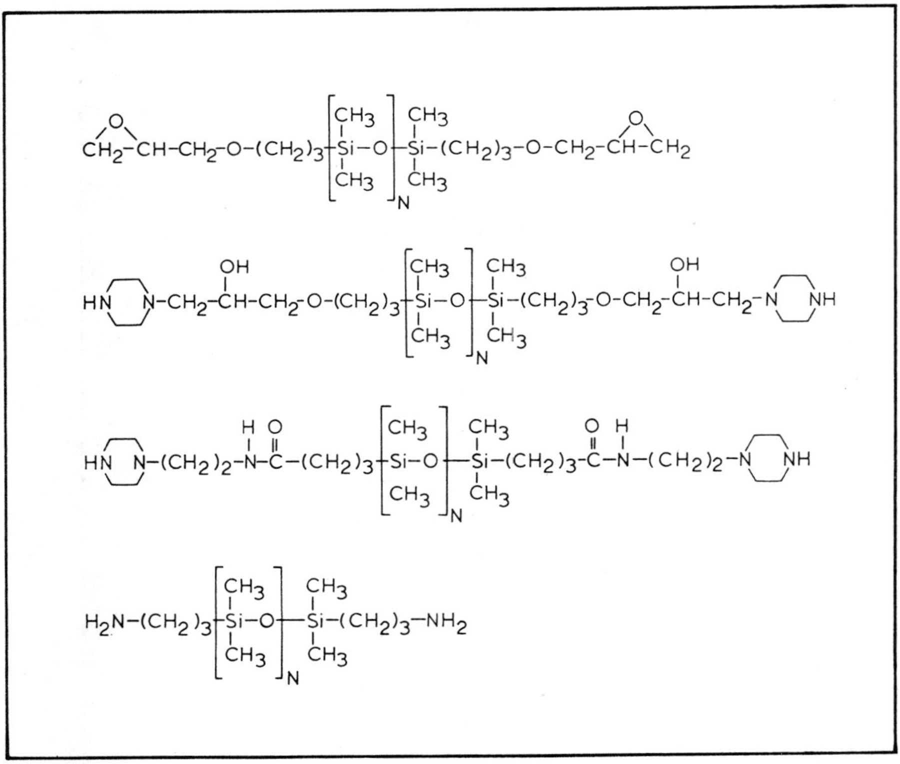

Scheme 9

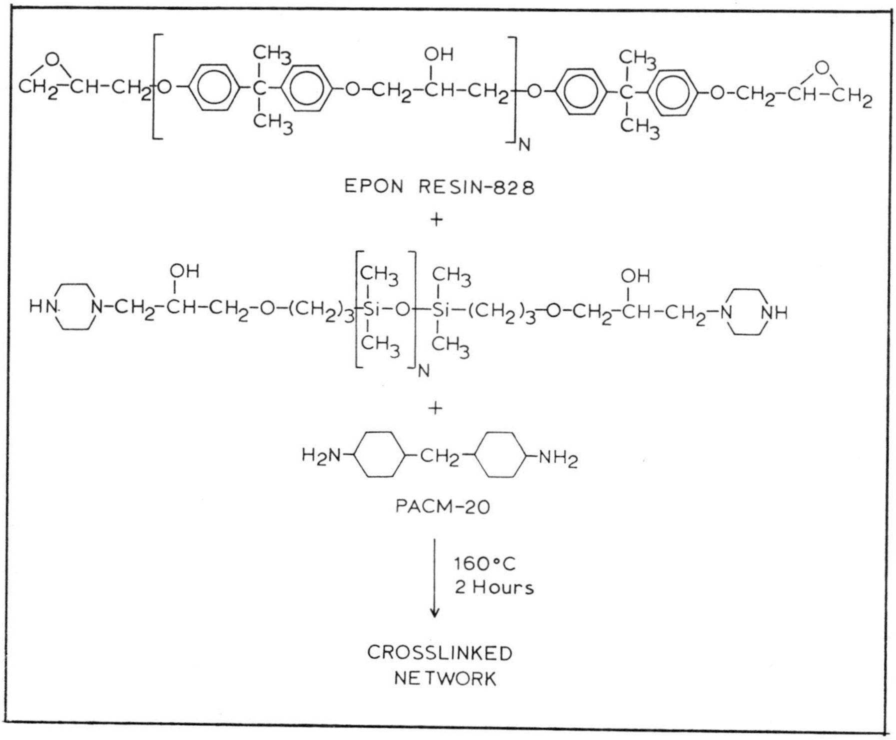

Scheme 10

piperazine capped diepoxy systems. Reactions were, in general, conducted for two hours at 160°C to produce a stable, well crosslinked network. The concentration of the modifiers was varied. The curing reaction was studied with the various siloxane modifiers already mentioned in Scheme 9. However, here it is important to note the question of miscibility. In order for the modifier to be effective, one imagines that it must first be reasonably miscible with the epoxy oligomer and subsequently perhaps microphase separate as small domains during the progress of the curing reaction itself. In the work described in the literature (10) utilizing butadiene/acrylonitrile reactive liquid polymer, one presumably incorporates acrylonitrile into the system to obtain an appropriate solubility parameter such that the liquid rubber will be miscible with the epoxy precursor, such as EPON Resin-828. In the case of siloxanes, the situation is somewhat more difficult. Pure polydimethylsiloxane has a calculated solubility parameter of about 7.6, which is much lower than that of the epoxy oligomers. Thus, when we tried to mix the epoxy terminated siloxanes with the EPON resin-828, it was quickly observed that the systems were highly incompatible (immiscible). If one attempts to prepare a cured specimen from the epoxy capped siloxanes of this type, one finds that the resulting product is extremely heterogeneous in a non-controllable way and that indeed unreacted liquid siloxane remains in the otherwise cured specimen after the time cycle mentioned. The aminopropyl siloxane oligomers appeared to be somewhat better but yet they were not highly miscible either. Additional problems with inhomogeneities were observed in this system. Moreover, in this case, the primary amine group on the siloxane modifier is capable itself of prematurely gelling the epoxy system. In order to circumvent these problems, it occurred to us that secondary amine systems might be of considerable interest. These modifiers can be prereacted with the EPON resin-828 since they should only react under moderate conditions with one epoxy group, it should be possible to prepare linear soluble precursor. Indeed, the reaction shown in Scheme 10 utilizing the piperazine capped epoxy oligomers was much different than all that had been studied prior to that point. The materials in this case apparently react to form homogeneous mixtures of the siloxane modifier and the EPON resin-828. In general, this reaction of the modifier with the EPON resin-828 was allowed to proceed for 2 hours at 60°C prior to introduction of the curing agent (PACM-20). The curing agent was added with rapid stirring. The mixtures of the active components were then cast into a silicone mold as described in the experimental. The curing step was conducted for 2 hours at 160°C. This temperature was chosen since we were able to demonstrate, as will be discussed later, that the glass transition temperatures were no higher than 150°C. Therefore, one would expect the reaction temperature would be sufficient to provide adequate mobility for the chain ends to react and generate the network structure. The other modifiers were reacted similarly to those in the above described procedure. In order to follow the reaction during the curing stage, we utilized differential scanning calorimetry. A variety of disiloxanes noted as "DSX" and polysiloxanes noted as "PSX" were utilized. In Figure 3 we show a comparison of the conversion temperature behavior for dynamic DSC scans utilizing a heating rate of 10°C/minute. Sample 3 in Figure 3 may be considered to be the control experiment utilizing the cycloaliphatic diamine as the curing agent. Sample 4 illustrates the reaction of the epoxy disiloxane with the standard curing agent, again PACM-20. One may note from this curve that the fastest reaction is obtained with the piperazine capped disiloxane system. All of these reactions were conducted in this case at 1:1 stoichiometry. The aminopropyl disiloxane also reacts quite fast although not as fast as the piperazine adduct. It is important to note that by far the slowest reaction was observed between the epoxy disiloxane and the PACM-20 curing agent. Additional information concerning the reaction rate of these various materials is provided in Figure 4. Here again the same designations that were noted in Figure 3 are applicable.

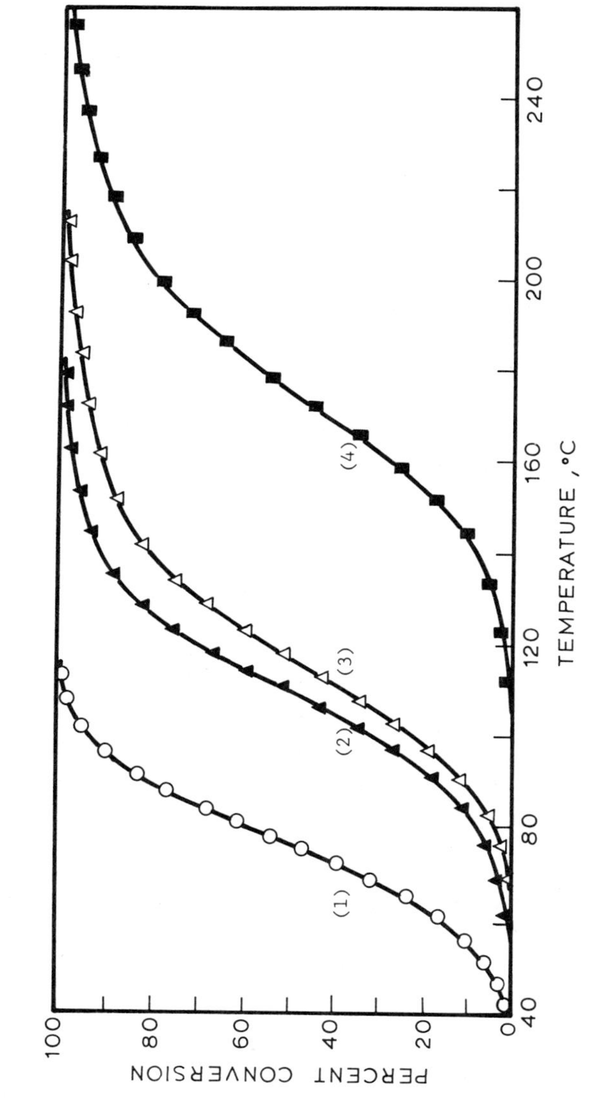

Figure 3. Comparison of Percent Conversion-Temperature Behavior of Various Siloxane Modified Epoxy-Amine Systems. Dynamic DSC Scans with a Heating Rate of 10°C/Minute.

(1) o EPON Resin-828 + Piperazine Capped Siloxane Dimer (DSX)

(2) ▲ EPON Resin-828 + Aminopropyl Terminated DSX

(3) △ EPON Resin-828 + PACM-20

(4) ■ Epoxy Terminated DSX + PACM-20

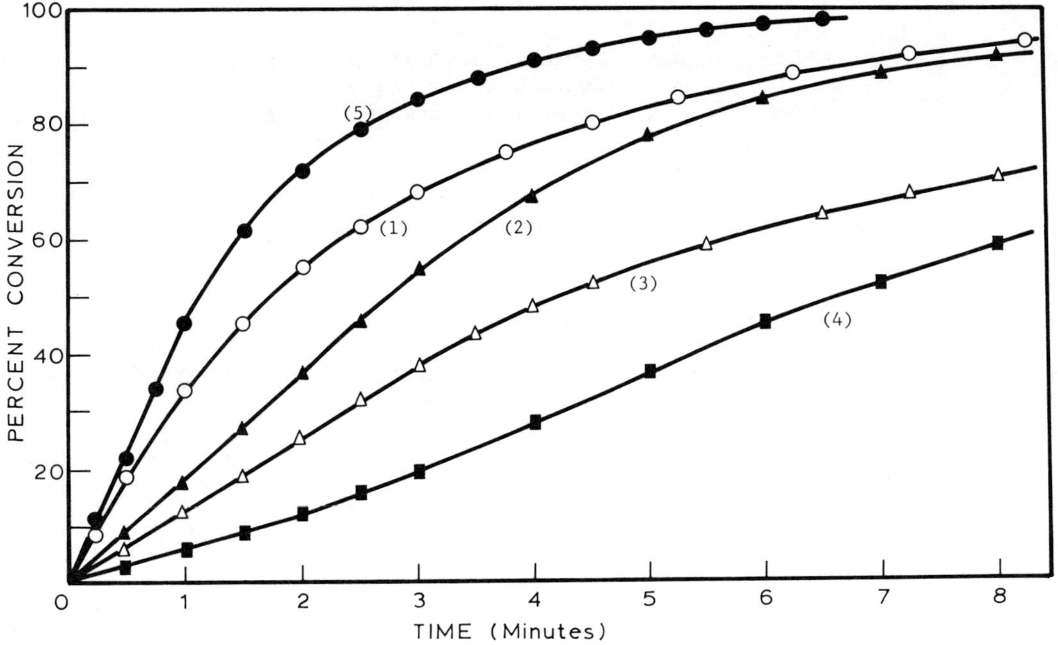

Figure 4. Comparison of Isothermal Reaction Rates (Reactivities) of Various S loxane Modifiers with EPON Resin-828 at 100°C.
(1) o EPON Resin-828 + Piperazine Capped Siloxane Dimer (DSX) (at 80°C)
(2) ▲ EPON Resin-828 + Aminopropyl Terminated DSX
(3) △ EPON Resin-828 + PACM-20
(4) ■ Epoxy Terminated DSX + PACM-20
(5) ● EPON Resin-828 + Piperazine Capped PSX ($\overline{M}n$=4200)

One observes again that, as a function of time, by far and away the fastest reaction is observed with the piperazine capped disiloxane. Indeed, even at 80°C, it is very rapid and was too fast to follow by DSC at 100°C. The piperazine capped siloxane oligomer reacts very rapidly at 100°C, even faster than the aminopropyldisiloxane at 100°C. The curing agent control, as shown in the dark triangles, is much slower at 100°C than the modifiers discussed above. Hence, the modifiers should be reacted or used up faster than the curing agent in the ultimate systems of interest. Again, as was shown in Figure 3, by far the slowest reaction of this series studied was the epoxy disiloxane reacted with the PACM-20 curing agent. Its reaction, even at 140°C, was very slow. This no doubt at least partially explains why attempts to incorporate the epoxy capped siloxane into the network were unsuccessful. That is, the reaction rate was so slow that most of the epoxy terminated oligomers were not incorporated, and hence, simply remained as incompatible additives after the curing reaction. The additional data derived from the DSC studies are shown in Table II. The reaction enthalpies were derived and appear to be in a reasonable region for this type of reaction.

Representative glass temperatures for a control and for modified systems prepared during this study are shown in Table III. In general, the glass temperatures of the PACM-20/EPON resin-828 system appear to be approximately 150°C. The effect of the modifier on the glass temperature of the network is very markedly dependent upon the length of the siloxane system. This point will be further addressed in future papers. Disiloxanes can in some cases plasticize the network whereas longer chains appear to develop a second microphase and therefore have only a nominal depression of the principal glass transition temperature.

Surface Analysis of Siloxane Modified Networks

ESCA

Electron spectroscopy has been utilized by a number of investigators to study the surface behavior of block copolymers and block copolymer/homopolymer mixtures (17-19,24,25). Generally the ESCA experiment is expected to allow one to ascertain the surface structure of perhaps the top 40 or 50 Å near the surface (20,21). We have thus far investigated the possible surface segregation of the cured epoxy systems by ESCA using three different modifiers, the epoxy dimer, the aminopropyl dimer and the hydroxypiperazine dimer. In addition, we also utilized oligomers based on the epoxy systems and also the hydroxypiperazine capped system. The data we have obtained is summarized in Table IV. Four low levels of siloxane were investigated ranging from 0.1% up to 2% by weight. Let us focus our attention first on the three dimers. By assessing the ratio

Table II

HEATS OF REACTION FOR THE CURING OF

EPOXY-AMINE SYSTEMS AS OBTAINED BY DSC

SYSTEM	ΔH_{Rxn} (kCal/eq)
EPON 828* + 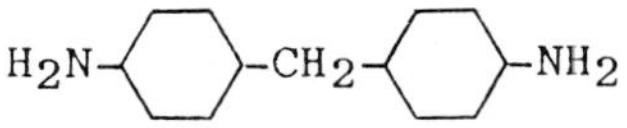	22.8 ± 1.3
EPON 828* + H_2N-DSX-NH_2	25.4 ± 1.2
EPON 828* + 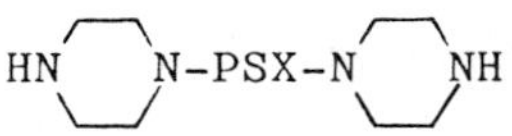 (a)	19.3 ± 1.5
EPON 828* + 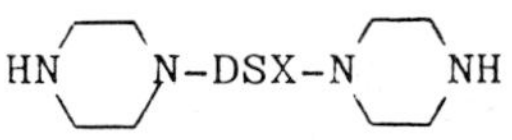 (b)	24.4 ± 1.1
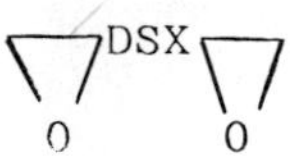 + 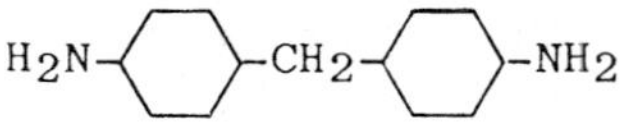	23.8

*EPON 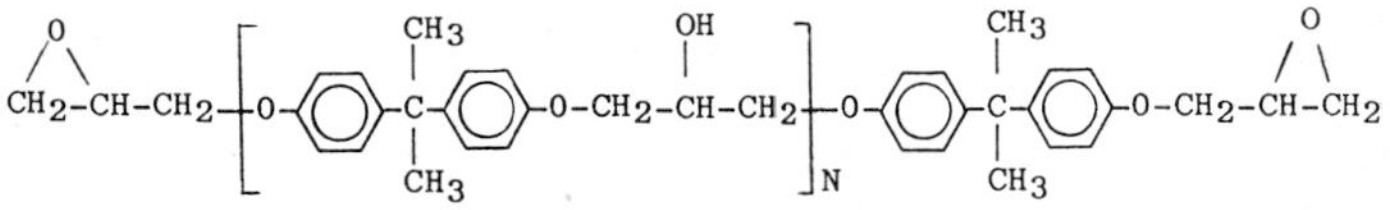

(a) Piperazine Capped Epoxy Functional Oligomeric Polydimethylsiloxane (M_n = 4,200)

(b) Piperazine Capped Epoxy Disiloxane

Table III

GLASS TRANSITION TEMPERATURES OF SILOXANE MODIFIED EPOXY NETWORKS [a]

Sample #	Siloxane Modifier Type	MW	% Wt	Tg (°C)
1	Aminopropyl	248.5[b]	4.0	127
2	Aminopropyl	248.5	10.0	113
3	Aminopropyl	2000	6.5	148
4	AEP[c]	850	10.0	146
5	AEP	2750	10.0	150
6	PIP.[d]	4200	8.0	147
7	PIP.	4200	10.0	149
8	CONTROL (EPON Resin-828 + PACM-20)			150

(a) DSC, 10°K/Minute

(b) Dimer

(c) Aminoethylpiperazine capped PSX

(d) Piperazine capped PSX

Table IV

ESCA C/Si RATIOS OF VARIOUS FUNCTIONAL DIMER AND OLIGOMERIC POLYDIMETHYLSILOXANE MODIFIED EPOXY NETWORKS (a)

WT. % SILOXANE	EPOXY DIMER	AMINOPROPYL DIMER	PIPERAZINE* DIMER	EPOXY** OLIGOMER	PIPERAZINE*** OLIGOMER
0.1	17.3	--	200	2.3	3.4
0.5	9.7	15.9	179	2.1	3.9
1.0	6.1	16.4	161	1.9	3.3
2.0	9.1	9.0	125	1.9	2.9

* 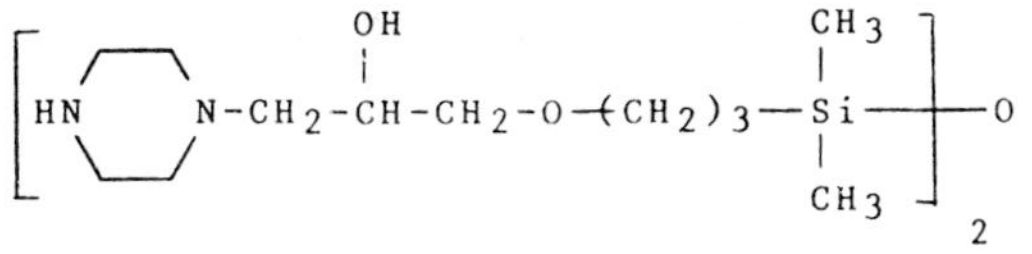

** 2900 g/mole

*** 4200 g/mole 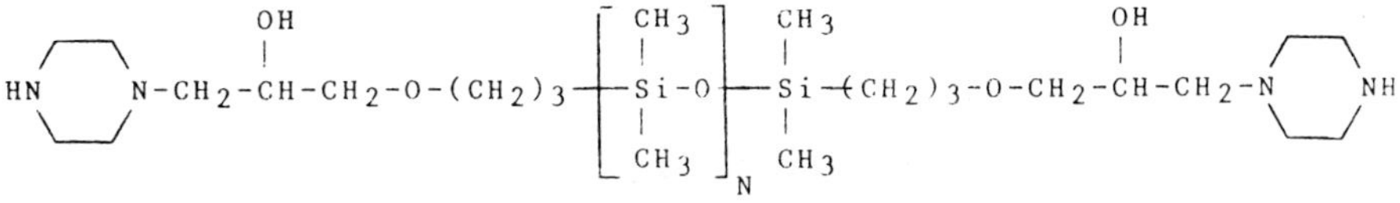

(a) A C/Si ratio of 2.0 would be indicative of a siloxane-like surface

of carbon to silicon as judged in this manner, if one had pure siloxane on the surface, the carbon to silicon ratio should be exactly 2.0. Clearly, in all of the dimers this is not the case, which indicates that for the dimeric modifiers significant amounts of epoxy resin-828 units are near the surface. However, it is appropriate to point out that the piperazine dimer is quite different from either the epoxy dimer or the aminopropyl dimer. The high ratios here imply that in fact very little of the siloxane dimer is on the surface in the case of the piperazine dimer. This is reasonable since the simple disiloxane is in fact rather polar. On the other hand, the highest concentration of siloxane is observed with the epoxy disiloxane. This can be interpreted as being due to the fact that the epoxy system is the most incompatible of the three units compared here. The two oligomers studied included the highly incompatible epoxy oligomer which was basically not miscible at all with the EPON resin-828. Here one observes rather close to 100% coverage of the surface with siloxane (100% coverage would be a value of 2.0). In the case of the piperazine capped oligomer of about 4,000 molecular weight, it is clear that the surface is rather siloxane rich as judged by the relatively low C/Si values. However, under these conditions, it is not quantitatively covered by the siloxane. Nevertheless, one may observe significant physical differences in these systems as compared to the unmodified epoxy materials. They are indeed low energy surfaces even at this ratio. Additional work on this question of surface segregation using ESCA is being continued.

Some additional studies of the surface enrichment phenomenon were invesigated by using an ATR-FTIR study. In this case, we utilized siloxane modified epoxies, prepared by reacting the piperazine capped epoxy dimer or oligomer with EPON resin-828 and PACM-20. The infrared spectra of a network incorporating 2% of a piperazine oligomer as shown in Figure 5 may be contrasted with an analogous network with 2% piperazine capped siloxane dimer (Figure 6). The peak due to the siloxane at about 1011 or 1012 cm^{-1} is much stronger in the piperazine oligomer than it is in the piperazine disiloxane system. This information is in agreement then with the ESCA data that showed the dimer to be much less capable of enriching the surface than the oligomer. Additional studies on the ATR modified systems are in progress for other oligomers in this series as well. It it clear however that both surface probes are in agreement. Also, the qualitative observations of the surface "feel" indicate that one has a great deal of enrichment of a siloxane on the surface relative to the concentrations that are present in the bulk. Additional quantitative work is being pursued with higher resolution instruments.

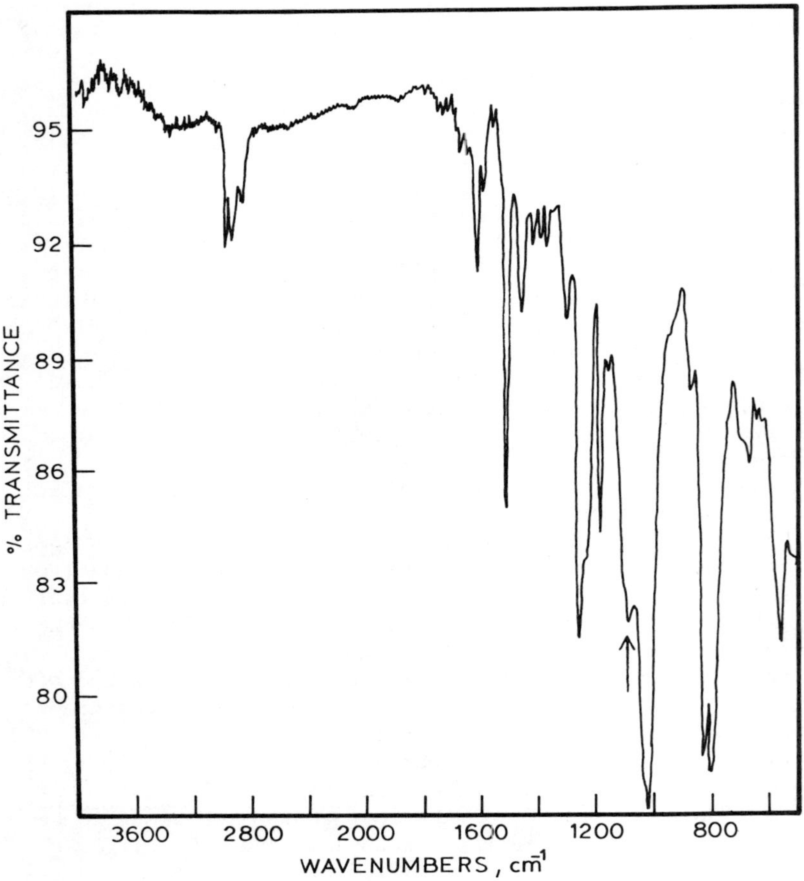

Figure 5. ATR/FTIR Spectrum of 2% Piperazine Terminated Siloxane Dimer-Modified Epoxy Networks.

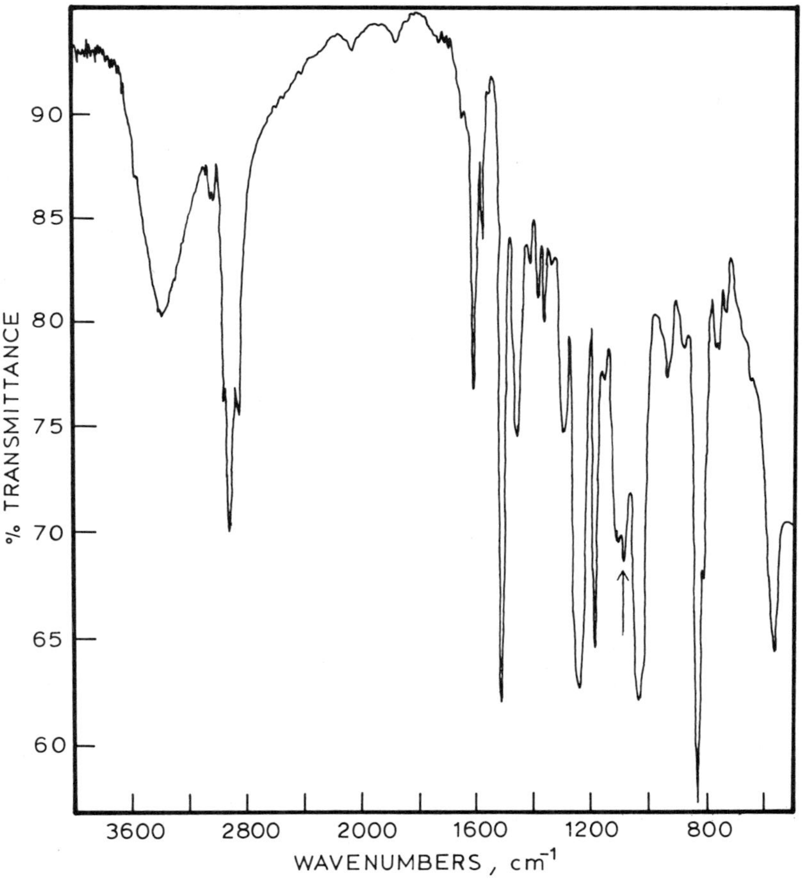

Figure 6. ATR/FTIR Spectrum of 2% Piperazine Terminated Siloxane Oligomer Modified Epoxy Networks ($\bar{M}n$ = 4200).

Synthesis of Linear Copolymers

It is of interest to investigate whether linear soluble copolymers could be prepared from these systems. To do this, we utilized a reaction sequence shown in Scheme 11. The reaction was conducted by dissolving the epoxy capped siloxane oligomer and the bisphenol-A diglycidyl ether in dioxane to form about a 50% solution. The reaction was heated to 60°C and a stoichiometric quantity of piperazine was added. The reaction was allowed to proceed for up to 48 hours and it was observed that the viscosity became rather high after 24 hours. In view of the reaction kinetics discussed earlier, it seems probable that most of the piperazine first reacts with the diglycidyl ether and later, the piperazine capped diglycidyl ether further reacts with the epoxy functional siloxane to produce a linear copolymer. In any event, it was possible to make a reaction product after 48 hours and to cast clear transparent films that adhered very strongly to glass surfaces. Further work on these segmented linear polyhydroxyether siloxane copolymers is continuing. Indeed, such materials may be of interest in their own right as novel adhesives.

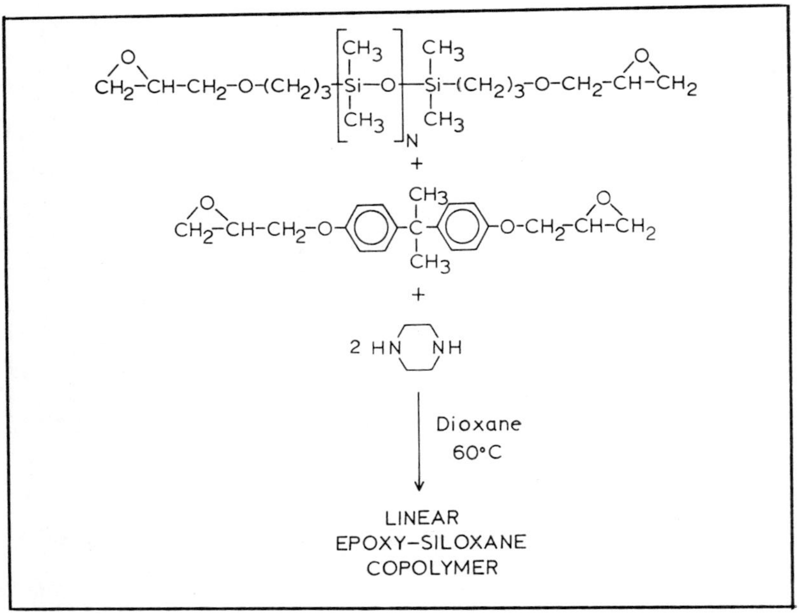

Scheme 11

CONCLUSIONS

Functional siloxane oligomers bearing either epoxy, aminopropyl, hydroxy piperazine or piperazine endgroups were synthesized via equilibration polymerization of the cyclic tetramer and the appropriate disiloxane. The reactions were further characterized by spectroscopic analysis and in some cases gel permeation chromatography. Incorporation of the siloxanes into epoxy networks was investigated. The behavior of the resulting networks was highly dependent upon the nature of the functional endgroup. The reactivity of the epoxy functional siloxane systems was very slow relative to the curing reaction and hence, epoxy functional siloxanes produced undesirable, turbid, incompatible films. Aminopropyl terminated modifiers also produced inhomogeneous networks. Only networks prepared with the secondary amine terminated systems, e.g. piperazine functionality, were able to produce homogeneous cured networks. Networks of starting material disiloxanes were also investigated and in those cases it was possible to produce transparent somewhat tough films. However, the glass temperature was reduced by the short siloxane material. Oligomeric species up to 5000 molecular weight, however, did not significantly depress the glass temperature of the network from its control value of about 150°C. Surface modification of the epoxy networks was studied by both ESCA spectroscopy and ATR-FTIR methods. In both cases, it could be demonstrated that piperazine capped oligomers were able to significantly enrich the surface with siloxane structures.

FUTURE WORK

We have at this point generated functional oligomers that can be quantitatively incorporated into epoxy networks. It is clear that surface modification is possible by this method. Our future work is focusing on the mechanical property behavior of these modified siloxane systems. In particular, we are investigating the morphology and structure of the dispersed phase as a function of siloxane concentration and block length. Stress strain properties at low extension rates as well as impact studies using our recently designed impact tester are being pursued (22).

Acknowledgment

The authors would like to express their appreciation to the Office of Naval Research for funding the work described in this paper. They would also like to thank the Shell Development Company and The DuPont Company for supplying generous samples of EPON resin-828 and PACM-20, respectively.

Literature Cited

1. Potter, W. G., Epoxide Resins, Springer-Verlag, N.Y., 1970.
2. Bauer, R. S., Ed., Epoxy Resin Chemistry, ACS Symp. Ser. #114, Wash. D.C., 1979.

3. May, C. A. and Tanaka, Y., Eds., Epoxy Resin Chemistry and Technology, Marcel Dekker, N.Y., 1973.
4. Lee, H., and Neville, K., Handbook of Epoxy Resins, McGraw-Hill, N.Y., 1967.
5. Sultan, J. N. and McGarry, F., Polym. Eng. Sci., 13, 29 (1973).
6. Rowe, E. H., Siebert, A. R., and Drake, R. S., Mod. Plast., 49, 110 (Aug., 1970); Rien, C. K, Rowe, E. H. and Siebert, A. R., "Toughness and Brittleness of Plastics", Deanin, R., and Crugnola, A. M., Editors, Adv. in Chem., No. 154 1974, p. 326-343.
7. Meeks, A. C. Polymer, 15, 675 (1974).
8. Soldatos, A. C., and Burhans, A. S., ACS Adv. Chem. Ser., 99, 531 (1971).
9. Bascom, W. D., and Cottington, R. L., J. Adhesion, 7, 333 (1976).
10. Riew, C. K., Rubber Chem. and Tech., 54(2), 374 (1981).
11. Drake, R. S., Egan, D. R., and Murphy, W. T., Organic Coatings and Applied Polym. Sci. Prep., 46, 392 (1982).
12. Arkles, B. C., Peterson, W. R., and Anderson, R., eds., Silicon Compounds: Register and Review, 2nd ed., Petrarch, Bristol, Pa., 1982.
13. Gilbert, A. K., and Kantor, S. W., J. Polym. Sci., 40, 35 (1959).
14. Lunsford, D. J., Synthesis and Characterization of Polyhydroxyethers, M. S. Thesis, V.P.I. & S.U., 1981.
15. Turi, E. A., ed., Thermal Characterization of Polymeric Materials, Academic Press, N.Y., 1981.
16. Yilgör, I., Yilgör, E., Banthia, A. K., Wilkes, G. L., and McGrath, J. E., Polymer Bulletin, 4, 323 (1981).
17. Clark, D. T., and Reeling, J., J. Polym. Sci., Polym. Chem. Ed., 14, 543 (1976).
18. Clark, D. T., Dilks, A., Peeling, J., and Thomas, H. R., Faraday Disc., Chem. Soc., 60, 183 (1976).
19. Thomas, H. R., and O'Malley, J. J., Macromolecules, 12(2), 323 (1979).
20. Clark, D. T., and Thomas, H. R., J. Polym. Sci., Polym. Chem., Ed., 15, 2843 (1977).
21. Clark, D. T., in Adv. Polym. Sci., H. J. Cantow, Ed., Springer-Verlag, Berlin, 125 (1977).
22. Wnuk, A. J., Ph.D. Thesis, V.P.I. & S.U., Blacksburg, Va., August, 1979; Wnuk, A. J, Ward, T. C., and McGrath, J. E., Polym. Eng. Sci., 21(6), 313 (1981).
23. Haward, R. N., Editor, The Physics of Glassy Polymers, Wiley (1973).
24. Gaines, G. L., Macromolecules, 14, 208 (1981).
25. McGrath, J. E., et al., Polymer Preprints, 20(2), 528 (1979).

RECEIVED March 2, 1983

3

Synthesis and Analysis of Saturated, Reactive *n*-Butyl Acrylate Polymers for Use in Epoxy Resin Toughening

S. GAZIT[1] and JAMES P. BELL

University of Connecticut, Storrs, CT 06268

A low molecular weight, carboxyl terminated poly (n-butyl acrylate) rubber was synthesized for evaluation as a toughening agent in epoxy resins. The carboxyl ends were incorporated by initiation with azo bis-cyanovaleric acid and by use of dithiodiglycolic acid as a chain transfer agent to control molecular weight. Bulk polymerization was required to avoid severe chain transfer to solvents, thus providing functionality near 2.0. The molecular weight decreased with increasing chain transfer agent, initiator concentration, and with increasing polymerization temperature.

Epoxy resins with relatively high glass transition temperature (>100°c) are brittle, Lee and Neville (1) Perez (6). The toughening of crosslinked epoxy resins, such as the diglycidyl ether of bisphenol A (DGEBA) by Reactive Liquid Polymers (RLP) has been the subject of intensive studies since the mid-sixties, Bucknall (7). One method of toughening epoxy by RLP is based on incorporating a small proportion of a low glass transition polymer, typically between 5 and 20%, into the rigid matrix. Initially, the rubber is miscible with the mixture of the epoxy monomer and the curing agent. During the curing process most of the RLP precipitates from the matrix and forms a homogeneously dispersed, rubbery, fine particle phase. This two phase morphology improves the toughness of the host matrix, Manson and Sperling (3). If the rubbery polymer fails to separate from the epoxy matrix it serves as a plasticizer rather than a toughening element, and while toughness may be improved, there is a major reduction of the sofening temperature. The success of this principle with

[1] Current address: Rogers Corporation, Lurie Research and Development Center, Rogers, CT 06263

0097-6156/83/0221-0055$06.00/0

presently available DGEBA/ RLP systems has not yet reached the levels observed with rubber toughened thermoplastics such as polystyrene.

The toughening of epoxy resins has been attempted with various rubbers. Noshay and Robeson (8) copolymerized DGEBA with low molecular weight polycaprolactone and polypropylene oxide. Sperling and Friedman (9) introduced the use of Simultaneous Interpenetrating Network (SIN) to improve the impact strength of epoxy. In this method monomers of DGEBA and n-butyl acrylate were simultaneously polymerized in bulk. McGarry and Willner (10) studied the interaction between DGEBA and butadiene based various Hycar rubbery resins. Most of the recent studies in the field concentrated on these Carboxyl Terminated Butadiene-Acrylo-Nitrile (CTBN) rubbers. A few other liquid, linear reactive modifiers have been proposed recently to toughen epoxy resins. Carboxyl-terminated polyisobutylene (CTPIB) was synthesized and studied by Slysh (11). United Technologies Corp. has also commercialized ABAN (carboxylated butadiene/ acrylonitrile copolymers) for toughening DGEBA.

It is well known that for effective toughening of the epoxy resin the rubbery elements have to be grafted, to a certain degree, to the DGEBA matrix (10,12,13). However, an excessive amount of chemical interaction between the DGEBA and the rubber modifier might lead to the formation of a single phase, undesirable morphology (8).

Various reactive groups have been used for chemical grafting between the epoxy resin and the rubbery phase; carboxyl (10,11), hydroxyl (14), epoxy (13,14), mercaptan (14) and amine (ATBN, a Hycar resin). The reactive groups can be present randomly on the rubber molecule backbone or can be located at the ends of the polymeric chain. Most researchers suggest that polymers possessing two reactive terminal groups (telechelic polymers) produce stronger elastomeric structures than randomly functional RLPs with the same molecular weight (10,11,18).

A new RLP, Carboxyl Terminated Poly n-Butyl Acrylate (CTPnBA), reported in this study, falls into the category of telechelic polymers, possessing on the average 1.8-2.0 carboxyl terminal groups per chain. The CTPnBA has theoretically a high potential for toughening DGEBA for the following reasons:

Physical--1. The glass transition temperature of PnBA occurs between -40°C and -50°C. It has a relatively high mechanical damping peak which is helpful for the toughening mechanism of DGEBA/PnBA composites.

2. At room temperature PnBA (up to $\bar{M}n \approx 25,000$) is a viscous liquid.

3. Low molecular weight liquid PnBA is soluble in DGEBA at moderate temperatures. The difference in the solubility parameters of PnBA and DGEBA is relatively small (2).

Chemical--4. PnBA is more stable chemically than the butadiene based RLPs, which contain residual double bonds.

5. The nBA liquid monomer (b.p. 145°C) is polymerized readily, using relatively convenient conditions.

Experimental

The Carboxyl Terminated Poly n-Butyl Acrylate was polymerized in a glass, 500 ml. batch reactor. The reactor was designed for solution and bulk polymerization under atmospheric pressure and variable temperatures, Figure 1. In the solution polymerization tests the initiator, 4,4'-azo bis-(4-cyanovaleric acid) (ABCVA), and the chain transfer agent, dithiodiglycolic acid (DTDGA), were dissolved in the polymerization solvent. The monomers n-butyl acrylate (nBA) and ethylene diacrylate (EDA) were added to the polymerization media at a constant flow rate through a syringe pump arrangement. The reactor was purged with nitrogen at the beginning of the polymerization process, and a nitrogen blanket was maintained over the solution during the entire reaction period. In the bulk polymerization procedure the initiator and the chain transfer agent powders were thoroughly mixed and placed (dry) in the reactor. After the reactor was purged with nitrogen the nBA monomer was poured into the polymerization container at once, and rapidly mixed with the powders. The bulk polymerization was rapid and occurred within a few minutes between 69° and 120°c.

In the analytical procedure the CTPnBA was separated from the excess initiator, chain transfer agent and monomeric nBA in a cleaning method which included several stages, Appendix A. The molecular weight of the purified CTPnBA rubber was measured by a Knauer Vapor Pressure Osmometer in chloroform solution at 37°C. The carboxyl content of the CTPnBA was determined by a chemical titration procedure using a modified potentiometric technique, Appendix B.

Background Calculations

The kinetics of dilute solution polymerization of butyl acrylate were studied by Melville and Bickel (15). They measured the propagation and termination coefficients at 25°c; $k_p = 13$ (ℓ/mole sec) and $k_t = 1.8 \times 10^4$ (ℓ/mole sec). Melville and Bickel also concluded that the termination reaction at 25°C occurred mostly through combination, thus theoretically increasing the probability to produce a carboxyl terminated PnBA with an ideal functionality of f=2 when the polymerization reaction is initiated with ABCVA.

Calculation of the degree of polymerization of a free radical polymerization in very dilute solution, assuming chain transfer reactions and termination by combination, indicated $\bar{X}n = 40\text{-}70$.

Figure 1. The reactor assembly for the polymerization of PnBA. Key: a, mixer; b, condenser; c, syringe; d, syringe drive system; e, reactor; f, oil bath; g, vapor trap; and h, nitrogen flow meter.

This is equivalent to a molecular weight of $\bar{M}n$ = 6000-8000, Appendix C.

The enthalpy of mixing of two liquids is expressed by eq. 1, Hildebrand and Scott (4)

$$\Delta Hm = Vm\ [\delta_1 - \delta_2]^2\ \phi_1 \phi_2 \quad (1)$$

where δ is the solubility parameter, Vm is the molar volume of the mixture, and ϕ_1, ϕ_2 are volume fractions of solvent and polymer, respectively.

The solubility parameter δ may be calculated from the group contribution principle developed by Small (16). A more complete treatment takes into account secondary effects, Olabisi et al. (5). The enthalpy of mixing controls the miscibility of two liquid polymers; if $(\delta_1 - \delta_2)^2 = 0$, the solution is assured by the positive entropy term in the Gibbs Equation $\Delta Hm - Tm\Delta Sm$, making ΔG negative (favorable). The calculated solubility parameters of PnBA and DGEBA are $\delta_{PnBA} = 9.24\ (cal/cm^3)^{1/2}$ and $\delta_{DGEBA} = 9.70\ (cal/cm^3)^{1/2}$ respectively. This relatively small difference in the solubility parameters of the two resins theoretically favors the probability of the spontaneous mixing of PnBA and Epon 828. However, the miscibility is also dependent on the molecular weight of the resins--the higher the molecular weight of the PnBA the lower the δ, and thus the lower the solubility of the rubber in the epoxy resin.

Results and Discussion

Polymerization of CTPnBA in dilute solution. The solution polymerization of N-butyl acrylate is better predicted and simpler to control. The effect of the molecular weight on the miscibility of CTPnBA in Epon 828 was studied with a series of polymers polymerized in dilute t-butanol solutions, Table I. As expected, increasing both dilution of the monomer and the polymerization temperature resulted in decreasing molecular weight of CTPnBA. The solubility of 10-15 parts of CTPnBA in 100 parts DGEBA (Epon 828) at 25°C was the miscibility criterion for the two resins. The criterion was chosen since the toughening mechanism of CTPnBA/ Epon 828 systems was expected to be effective with this resin ratio range, based on CTBN rubber studies (10,14). The results suggested that the upper limiting molecular weight for the solubility of CTPnBA in Epon 828 was around $\bar{M}n$ = 4400.

The functionality of polymer 1-b was found to be surprisingly low f=1.15, apparently attributable to extensive chain transfer to the solvent or due to disproportionation termination. Tert-butanol was reported to be one of the few solvent systems in having an extremely low chain transfer potential in similar free radical polymerizations (17). The solubility of the reagents, the polymerization temperature and the reactivity characteristics restrict the choice of solvents to only a few, some of which were

TABLE I. Polymerization of nBA in t-butanol

#	t-butanol (cc)	nBA (g)	nBA (cc/min)+	ABCVA (g)	Temp. (°c)	$\bar{M}n$	Solubility*
1-a	450	15.75	0.12	1.71	70	6400	No
1-b	500	25.00	0.17	1.71	83	4400	Yes
1-c	1000	37.25	0.15	3.42	83	3850	Yes

* solubility of 10 parts of CTPnBA in 100 parts DGEBA.
\+ flow rate of nBA into the reactor.

TABLE II. Functionality of PnBA in various solvents.

#	Solvent type	Solvent (cc)	nBA (g)	nBA (cc/min)	ABCVA (g)	Temp. (°C)	$\bar{M}n$	f*
2-a	t-butanol	500	25.0	0.17	1.71	83	4400	1.15
2-b	acetone	500	25.0	0.19	1.71	56	6900	1.30
2-c	THF	500	14.0	0.34	1.71	65	1350	0.30

* f - functionality

studied, Table II. The functionality of CTPnBA was higher in acetone than in t-butanol, either due to a different extent of chain transferring to the solvents, or due to a different ratio of the termination by combination vs. disproportionation of the polymerization in the two solvents. The molecular weight of polymer 2-b (mainly attributable to low polymerization temperature, limited by the boiling point) was too high.

Efficient toughening by the rubber modifier can be obtained, as discussed earlier, only if the RLP is capable of sufficient grafting to the epoxy matrix. Theoretically, the optimal linear rubber chain should possess two terminal reactive groups. Since this configuration could not be synthesized in dilute solution polymerization, a branched copolymer of nBA and ethylene diacrylate (EDA), structure I, with a potentially higher functionality, was studied. Ethylene diacrylate is used primarily as a cross-

$$CH_2 = CH - \overset{\overset{O}{\|}}{C} - O - CH_2 - CH_2 - O - \overset{\overset{O}{\|}}{C} - CH = CH_2$$

I Ethylene diacrylate

linking agent in free radical polymerizations, but in this study it was used as a branching agent. The copolymerization of EDA with nBA was designed to build branching into the PnBA chain, and thus increase the probability of a functionality higher than 2. The additional carboxyl groups through this mechanism can be expected to be located at the tips of the branches. The functionality of the nBA/EDA copolymers increased with the concentration of EDA (Table III), but even this technique did not yield a sufficient increase in the functionality of the PnBA. The difference in functionality of 3-b and 3-c is insignificant and it indicates that increase in EDA concentration above the 3-b formulation does not lead to an increase in functionality.

TABLE III. Molecular weight and functionality of nBA/EDA copolymers.

#	t-butanol (cc)	nBA (g)	nBA (cc/min)	EDA (g)	ABCVA (g)	$\bar{M}n$	f
3-a	500	25	0.23	1.35	1.11	6100	1.15
3-b	500	25	0.23	3.40	1.11	6250	1.27
3-c	500	25	0.42	6.81	1.65	6200	1.30

Athey, Mosher and Weston (19, 20) studied the use of symmetrical difunctional chain transfer agents for synthesis of telechelic polymers. Dithiodiglycolic acid (DTDGA, II) was selected in the present work as a chain transfer agent to increase the functionality of the PnBA. This particular chain transfer agent cleaves at the sulphur-sulphur bond in the presence of a free radical as shown below:

$$P\cdot + OH-\overset{\overset{O}{\|}}{C}-CH_2-S-S-CH_2-\overset{\overset{O}{\|}}{C}-OH \rightarrow -PSCH_2-\overset{\overset{O}{\|}}{C}-OH + OH-\overset{\overset{O}{\|}}{C}-CH_2-S\cdot \qquad (2)$$

II Dithiodiglycolic acid

This mechanism increases the probability of obtaining an overall higher functionality of PnBA by scavanging the free radicals of the growing rubber chain with carboxyl carrier species. The results of the chain transfer study are summarized in Table IV. The effects of the concentration of DTDGA on the molecular weight of the polymer were dramatic, however the functionality of the PnBA still did not exceed f = 1.30.

TABLE IV. Molecular weight and functionality of CTPnBA, effect of DTDGA concentration.

#	t-butanol (cc)	nBA* (g)	ABCVA (g)	DTDGA (g)	$\bar{M}n$	f
4-a	450	16.2	1.71	0	13200	1.15
4-b	500	24.7	1.65	0.093	10100	/
4-c	500	25.6	1.65	0.468	4200	1.23
4-d	500	25.6	1.65	0.936	3800	1.30
4-e	500	25.2	1.65	1.174	3850	1.29

* The monomer was added to the solution at the beginning of the polymerization.

The results of the dilute solution polymerization suggested that solvent chain transfer was a governing factor in the termination step. By the same token, the mechanism of solvent chain transfer termination forms free radicals which do not carry carboxyl groups, and serve as potent initiator species in the polymerization media. It was concluded that in order to obtain higher functionality of the PnBA the solvent had to be eliminated from the polymerization reactor.

Polymerization of PnBA in bulk. The molecular weight of PnBA is very dependent on the polymerization temperature, in both bulk and solution polymerization. The effect of the polymerization temperature in bulk on the molecular weight of PnBA was studied from 69°C to 120°C, Table V. The molecular weight decreased dramatically with increasing polymerization temperature up to about 105-110°C, where it reached a plateau. Due to high molecular weight, none of the polymers was miscible with Epon 828 at room temperature.

The effect of the concentration of DTDGA on the molecular weight of CTPnBA was studied at a constant polymerization temperature of T = 108°C, Table 6. The results indicate that an increase in the concentration of the chain transfer agent beyond 10% mole ratio of DTDGA to nBA may reduce the molecular weight of the polymer by only a very small amount.

TABLE V. Molecular weight and functionality of CTPnBA at various polymerization temperatures, in bulk polymerization.

#	nBA (g)	ABCVA (g)	DTDGA (g)	Temp. (°C)	$\bar{M}n$	f
5-a	42.0	3.90	2.20	69	13500	/
5-b	42.0	3.90	2.20	80	12200	/
5-c	41.9	3.90	2.20	90	9700	2.06
5-d	42.0	3.89	2.20	100	7500	1.80
5-e	42.0	3.90	2.20	110	7000	1.78
5-f	42.1	3.91	2.20	120	6900	1.82

TABLE VI. Molecular weight and functionality of CTPnBA at various chain transfer concentrations in bulk polymerization.

#	nBA (g)	ABCVA (g)	DTDGA (g)	$\bar{M}n$	f
6-a	40.3	3.90	/	10300	2.16
6-b	39.9	3.91	1.01	8500	2.04
6-c	40.9	3.89	3.02	7000	1.96
6-d	40.8	3.88	5.99	6300	1.91

The effect of the initiator concentration on the molecular weight of the CTPnBA was studied at polymerization temperature of 120°C in the absence of the chain transfer agent, Table VII. As expected, increase in initiator concentration resulted in a decrease of the molecular weight.

The measured functionality of the CTPnBA rubber in the bulk polymerization was in the range of f = 1.80-2.16 (Tables V-VII). This value was reproducible in the various experiments and occurred independent of changes in polymerization temperature and concentrations of the reagents.

The lowest molecular weight of the CTPnBA polymerized in bulk was in the range of $\bar{M}n$ = 6000-7000. Carboxyl Terminated Poly n-Butyl Acrylate with this high molecular weight was not soluble in Epon 828 at room temperature, but became miscible with the epoxy resin at 120°C. The addition of some amine curing agents also enhances solubility; a slightly cloudy mixture of the rubber in Epon 828 becomes clear when an aliphatic amine is added, Gazit (18).

TABLE VII. Molecular weight and functionality of CTPnBA at various initiator concentrations in bulk polymerization.

#	nBA (g)	ABCVA (g)	$\bar{M}n$	f
7-a	40.5	4.01	11900	1.93
7-b	40.6	5.00	11200	2.03
7-c	40.3	6.00	10000	1.95
7-d	40.2	7.01	9700	1.90

Conclusions

1. The molecular weight of CTPnBA is controlled by the polymerization temperature, initiator, and the chain transfer agent in both dilute solution and bulk polymerization.
2. CTPnBA rubber with molecular weight up to $\bar{M}n \sim 4400$ was miscible with Epon 828 at room temperature.
3. Severe chain transfer to solvent appeared to cause a limit in the functionality of CTPnBA to about f = 1.30 in dilute solution, free radical polymerization.
4. The polymerization temperature had the strongest effect on the molecular weight of CTPnBA. This was seen in both the solution and in the bulk polymerization experiments.
5. Increasing concentration of initiator and chain transfer agent resulted in decreasing molecular weight of the CTPnBA rubber.
6. Functionality of the CTPnBA rubber dramatically increased in the bulk polymerization experiments, suggesting that chain transfer to solvent was a significant reason for the poor functionality in dilute solution polymerization.

Appendix A

The CTPnBA cleaning procedure. The objective of the cleaning procedure was to remove the low molecular weight compounds from the CTPnBA rubber after the polymerization had been completed. The liquids were removed by evaporation under vacuum, and the solids were separated from the CTPnBA by extraction in solution.

(a) After evaporation of the polymerization solvent the mixture of CTPnBA and the low molecular weight compounds were dissolved in toluene.

(b) The ABCVA and DTDGA precipitated overnight were filtered several times until the solution became clear.

(c) The polymer solution in toluene was mixed with distilled water in a ratio of about 1 part of H_2O to 3 parts of toluene solution. The flask was shaken for about 1 hour, during which time the remaining dissolved solids in the toluene were extracted by the water. The organic polymer remained in the organic phase.

(d) The solution was centrifuged for about 1 hour at 2800 rpm, in a lab-bench centrifuge, and the two phases separated, the upper fraction containing the clean polymer in toluene.

(e) The toluene was evaporated in a rotary evaporator, and then placed in a vacuum oven at 80°C for about 12 hours until constant weight was obtained.

Steps (c) and (d) were repeated 3-4 times until the chemical titration showed no significant change in the acid content.

Appendix B

The determination of the carboxyl content in the CTPnBA. The method is based on a potentiometric acid-base titration technique in organic solution.

reagents = 1. Reagent absolute methanol
2. Reagent acetone
3. Standardized 0.5N KOH solution in methanol

An amount of 2-3 g of the cleaned and dried CTPnBA resin sample was titrated. The polymer sample was accurately weighed in a 300 cc round bottom flask. 50 cc of methanol and 100 cc of acetone were pipetted into the flask. The solution was stirred for about 20-30 minutes. The starting pH of the solution was in the range of 4-6. Standardized 0.5N KOH solution was then continuously added to the sample solution at a rate of about 0.1 cc/min by use of a syringe pump. The inflection endpoint occurred at about pH 10.5.

The acid-base titration procedure was checked with ABCVA and acrylic acid, and was found accurate and reproducible.

Functionality. The number of carboxyl equivalents was determined from the potentiometric acid-base titration. The number molecular weight $\bar{M}n$ of the CTPnBA was determined by the Vapor Pressure Osmometer measurement. The product of the number of acid equivalents by the molecular weight divided by the weight of the titrated PnBA sample is the calculated average number of carboxyl groups per chain of the poly n-butyl acrylate.

Appendix C

The calculation example of the degree of polymerization of n-butyl acrylate. The calculation assumes chain transfer to monomer, solvent and chain transfer agent for a typical batch.

monomer = nBA 30.0 gr, 33 cc (MW = 128)
initiator = ABCVA 1.7 gr, (MW = 280)
chain transfer agent = DTDGA 2.2 gr, (MW = 182)
solvent = t-butanol 393 gr, 500 cc (MW = 74)

The following rate constants were collected from Brandrup and Immergut (2), and Melville and Bickel (15), for the polymerization of n-butyl acrylate at 25°C.

$$C_M = 2.5 \times 10^{-5} \ (/)$$
$$C_S = 6.8 \times 10^{-6} \ (/) \ \text{(t-butanol)}$$
$$C_T = 6.5 \times 10^{-2} \ (/) \ \text{(DTDGA)}$$
$$kd = 1.55 \times 10^{-4} \ (1/\text{sec})$$
$$kp = 1.3 \times 10^{1} \ (\ell/\text{mol sec})$$
$$kt = 1.8 \times 10^{4} \ (\ell/\text{mol sec})$$
$$f = 0.6 \ \text{(assumed)}$$

It is assumed that there is a cancellation of the temperature effects on the rate constants, therefore they can be used for an estimated calculation of the degree of polymerization nBA at 80°C.

The rate of propagation is calculated by eq. C-1

$$R_p = kp \left(\frac{f\ kd\ [I]}{k_t}\right)^{1/2} [M] \qquad \text{(C-1)}$$

$$R_p = 4.3 \times 10^{-5} \ (\text{mole}/\ell \ \text{sec})$$

The degree of polymerization, $\bar{X}n = 36$, is calculated by the eq. below (Ref 21), and this value is equilvalent to an average molecular weight of $\bar{M}n = 4600$ for PnBA.

$$\frac{1}{\bar{X}_p} = \frac{k_t R_p}{K_p^2 [M]^2} + C_m + C_s \frac{[S]}{[M]} + C_t \frac{[T]}{[M]}$$

where R_p = rate of polymerization mols/ℓ-sec
R_t = termination rate const., ℓ/mol. sec.
k_p = propagation rate const. ℓ/mol. sec.
[M] = monomer concentration, mol/ℓ
[S] = solvent concentration, mol/ℓ
[T] = chain transfer agent concentration, mol/ℓ
C_m, C_s, C_t = chain transfer constants
$\bar{X}_n$ = degree of polymerization, number structural units/chain

Acknowledgment

The authors express their appreciation to Shell Development Company for support of this research.

Literature Cited

1. Lee, H.; Neville, K. "Handbook of Epoxy Resins"; McGraw-Hill, New York, 1967.
2. Brandrup, J.; Immergut, E. H. "Polymer Handbook"; John Wiley, New York, 1975.
3. Manson, J. A.; Sperling, L. H. "Polymer Blends and Composites"; Plenum Press, New York, 1976.
4. Hildebrand, J.; Scott, R. "The Solubility of Non Electrolytes"; Reinhold Publishing Corp., New York, 1949.
5. Olabisi, O.; Robeson, L. M.; Shaw, M. T. "Polymer-Polymer Miscibility"; Academic Press, New York, 1979, p. 52.
6. Perez, R. J. in "Epoxy Resin Technology"; Polymer Engineering and Technology, 1968, p. 45.
7. Bucknall, C. B. "Toughened Plastics"; Applied Science Publishers, London, 1977.
8. Noshay, A.; Robeson, L. H. J. Polymer Sci., Chem. Ed., 1974, 12, 689.
9. Sperling, L. H.; Friedman, D. W. J. Polymer Sci., part A-2, 1969, 7, 425.
10. McGarry, F. J.; Willner, A. M. "Toughening of an Epoxy Resin by an Elastomeric Second Phase"; R68-8, Massachusetts Institute of Technology, March, 1968.
11. Slysh, R. "Epoxy Resins"; ADVANCES IN CHEMISTRY SERIES No. 92, ACS: Washington, D. C., 1970; p. 108.
12. Bucknall, C. B.; Yoshii, T. The British Polym. J., 1978, 10, 53.
13. Scarito, P. R.; Sperling,L. H. Poly. Eng. and Sci., 1979, 19, 297.
14. Riew, C. K.; Rose, E. H.; Siebert, A. R. ADVANCES IN CHEMISTRY SERIES No. 154, ACS: Washington, D. C., 1976; p. 326.
15. Melville, H. W.; Bickel, A. F. Trans. Faraday Soc., 1949, 45, 1049.
16. Small, P. A. J. Appl. Chem. 1953, 3.
17. Siebert, A. R.; U. S. Patent, 3,285,949, 1966.
18. Gazit, S. "Toughening of Epoxy Resin by Acrylic Elastomer," Ph.D. Dissertation, University of Connecticut, Sept. 1980.
19. Athey, R. D.; Mosher, W. A.; Weston, N. W. J. Polym. Sci., 1977, 15, 1423.
20. Athey, R. D.; Mosher, W. A.; Weston, N. W. ACS Polymer Preprints, 20, No. 2, 1979.
21. Henrici-Olive, G.; Olive, S. Fortschr. Hochpolymer Forsch. 1961, 2, 496.

RECEIVED December 16, 1982

4

Impact Performance of Epoxy Resins with Poly(*n*-butyl acrylate) as the Reactive Liquid Rubber Modifier

S. GAZIT[1] and JAMES P. BELL

University of Connecticut, Chemical Engineering Department and Institute of Materials Science, Storrs, CT 06268

Poly n-butyl acrylate was used as the Reactive Liquid Rubber (RLP) modifier in the investigation of the toughening of epoxy resins of the diglycidyl ether of bisphenol A type. No plasticizing additives were used, so that the Tg remained above 100°c. Effective bonding between reactive groups on the ends of the rubber chains and the matrix was necessary for toughening. The impact strength was improved when the rubber was prereacted with the epoxy component before adding the curing agent. Solubility of the rubber in the epoxy resin before addition of curing agent was strongly dependent on rubber molecular weight, with a value of $\bar{M}n \approx 4000$ being desirable. The number and size of rubber particles was a function of the curing agent concentration; the average size was approximately 1 micron.

Under favorable conditions microcavitation (whitening), "butterfly pattern" formation and sometimes shear banding were observed. These were concurrent with an increase of up to 100% in tensile impact energy to break, largely attributable to increased ultimate elongation; the ultimate stress was similar or slightly less than controls. Notched Izod results gave a smaller improvement because of the deformation by rapid propagation of a crack from a defect. The tensile impact method gives deformation over a wider area, permitting the toughening mechanism to be effective.

[1] Current address: Rogers Corporation, Lurie Research and Development Center, Rogers, CT 06263

0097-6156/83/0221-0069$06.00/0

The impact strength, as well as most other physical properties of crosslinked epoxy resins, is controlled by the chemical structure and ratio of the epoxy resin and hardener, by any added fillers, and by the curing conditions used. Unfortunately, crosslinked glassy epoxy resins with relatively high glass transition temperatures (>100°c) are brittle, Lee and Neville (1) Perez (3). The poor impact strength of high glass transition epoxy resins is a crucial limiting factor for the usage of epoxies as structural materials and in composites.

The toughening of crosslinked DGEBA by Reactive Liquid Polymers (RLP) has been studied since about 1965. Several approaches have been tested using various RLP modifiers. Generally, the toughening method is based on the incorporation of a small proportion of a low Tg polymer, typically between 5% and 20%, into the rigid epoxy resin. On mixing, the RLP, the epoxy resin and the curing agent must be compatible. Then, during the curing cycle, elements of the RLP precipitate from the matrix and form a homogeneously dispersed particle phase, Manson and Sperling (2). It is believed that this particular two phase morphology is the key to the toughening mechanism of the host matrix (4-9). The success of the epoxy toughening principle by RLP depends vitally on the interaction of the rubber with the epoxy matrix, and on the phase separation mechanism. Control of these two major variables is complex and only partially understood (8,9). Experimental results of the mechanical properties of RLP modified epoxy resins are frequently conflicting, and depend on the preparation techniques of the polymer composites (9,10,11).

Several low molecular weight polymer modifiers have been used as RLPs with epoxy/hardener systems. Noshay and Robeson (12) copolymerized polycaprolactone and polypropylene oxide with epoxy resins. The curing of the epoxy in the presence of the modifier produced block copolymer structures of crosslinked epoxy and linear modifier segments. By this method the low molecular weight polymeric modifier flexibilized the epoxy/curing agent network, and drastically lowered the glass transition temperature of the crosslinked polymer. The use of higher molecular weight modifiers resulted in the formation of a two phase morphology. The Izod impact strength, which was used as the criterion for improved impact resistance, was only slightly better than the controls.

Sperling and Friedman (13) introduced the Simultaneous Interpenetrating Network (SIN) principle. In this method the DGEBA resin and monomers of n-butyl acrylate were simultaneously polymerized in bulk, one in the presence of the other. A two phase morphology emerged, with the rubber phase dispersed in the glass matrix. The SIN concept was further developed by adding small amounts of glycidyl methacrylate to the reactive monomers system, to improve the bond between the two polymeric networks,

Scarito and Sperling (14). The impact strength of the modified SIN system was indeed improved, but the glass transition temperature of the composite decreased substantially relative to the epoxy control.

The carboxyl terminated butadiene-acrylonitrile (CTBN)/DGEBA (Epon 828) system was introduced by McGarry and Willner (4). The compatible CTBN/Epon 828 mixture was cured in the presence of a Lewis base catalyst, DMP-30. Later, the basic RLP-epoxy resin system was modified and thoroughly studied with a variety of additives and curing agents (5,6,15-19). The impact strength of present CTBN/Epon 828 systems with Tg above 100°c is better than the impact strength observed with most other RLP modifiers. However, none of the presently RLP/DGEBA formulations has improved the impact strength of epoxy controls to the degree that has been demonstrated by RLP toughened thermoplastic polymers. The toughening mechanism of thermoset plastics with RLP is still not well understood.

The study of the toughening of DGEBA resin with a new RLP, Carboxyl Terminated n-Poly Butyl Acrylate (CTPnBA), is reported here. The CTPnBA rubber was synthesized by dilute solution polymerization and bulk polymerization techniques (20,21). The carboxyl functionality of the polymers varied from f = 1.3 (in solution polymerization) to f = 1.8-2.0 (in bulk polymerization).

Experimental

The epoxy networks reported in this study were from DGEBA type monomer (EPON 828 by Shell) and CTPnBA (synthesized as described by Gazit and Bell, 21). The molecular weight of the RLP was characterized by Vapor Pressure Osmometry (VPO) in chloroform solution at 37°c. The molecular weight of the CTPnBA varied between $\bar{M}n$ = 6600 and $\bar{M}n$ = 10,300. An acid/base potentiometric titration procedure in organic solution was used to determine the functionality of the CTPnBA rubber (20). The functionality of the CTPnBA polymerized in solution was relatively low, $f \leq 1.3$, while in the bulk polymerization of the functionality of the linear CTPnBA reached the theoretical maximum f = 2.0. One must be careful to completely remove any residual carboxyl-containing polymerization catalyst and chain transfer agent before functionality determination.

Three curing agents were studied, at various curing conditions: methylene dianiline (MDA), 2.5-dimethyl-2.5-hexanediamine (DMHDA), and DMP-30. The epoxy/RLP systems were cured between separated glass plates which had been treated with a fluorcarbon mold release agent; the hot (usually 80° to 120°c) liquid epoxy resin or epoxy/RLP composition was thoroughly hand mixed with the curing agent for about a minute. Air bubbles were removed from the mixture during a few minutes in a vacuum oven at 80°c. The clear mixture was then poured between the plates and placed in a controlled oven for the desired curing cycle. The cured plates were slowly cooled to room temperature after the curing cycle. Samples for the mechanical tests were machined to standard Notched Izod Test (ASTM D256-56), and to Tensile Impact Test dog-bone shapes.

The Tensile Impact Tester used in this study is a modified version of the Plastechon Universal Tester by Plas-Tech Equipment Co. It was operated at a ram speed of 1.5 m/sec, with a maximum stroke of 2.45 cm. Load was measured by a Dynisco TCFTS-1.5M solid-state bonded gage transducer. Displacement was measured by the Kaman Multi-Vit displacement measuring system, Model DK-2300-10CU. The Kaman transducer made non-contacting measurement of the diplacement possible. The readout of the two transducers was via an Explorer III digital oscilloscope by Nicolet Instrument Corporation. Explorer III is a two channel, high resolution, storage oscilloscope with a capability of storing signal waveforms on a magnetic disc. The oscilloscope recorded X/time and Y/time simultaneously and had the option of X/Y display. These features were used to construct stress-strain curves. Ultimate strain ε_u and stress σu were calculated directly from the oscilloscope readout. Impact strength was calculated by measuring the area under the stress-strain curve. Typical displacement/time and force/time curves obtained by this method are shown in Figure 1.

Scanning Electron Microscopy (SEM) with magnification capability of up to 20,000X was used for the study of the polymer composite fracture surfaces. The freshly fractured surfaces were coated with a 250Å thick Au-Pd conducting layer.

Dynamic Mechanical Spectroscopy was studied by Rheovibron, Model DDV-II at heating rate of 0.5°C/min. Standard samples were used; the Rheovibron was operated at 110 Hz.

Results and Discussion

Toughening of DGEBA by RLPs results from the two phase morphology formed by the controlled precipitation of rubbery particles from the epoxy/RLP/curing agent mixture during the curing process. This phase separation of the rubbery particles from the gelling matrix starts from a homogeneous solution of the system components. The miscibility of the Epon 828/CTPnBA systems depends on the concentrations of the components, the molecular weight of the rubber and the mixture temperature. A typical phase diagram of CTPnBA ($\bar{M}n$ = 10,300)/Epon 828 mixture is shown in Figure 2. The concentration/temperature relationship of Epon 828 and various CTPnBA systems can be described by a family of parallel curves, shifting from left to right with decreasing molecular weight of the CTPnBA (20). Since the optimal rubber concentration in the composite should be in the range of 10%-20%, the example CTPnBA shown in Figure 2 is not compatible with Epon 828 at room temperature and the desired concentration levels. In a separate study (21) it was concluded that the limiting molecular weight for the solubility of CTPnBA in Epon 828 at low concentrations ($\approx$ 10%

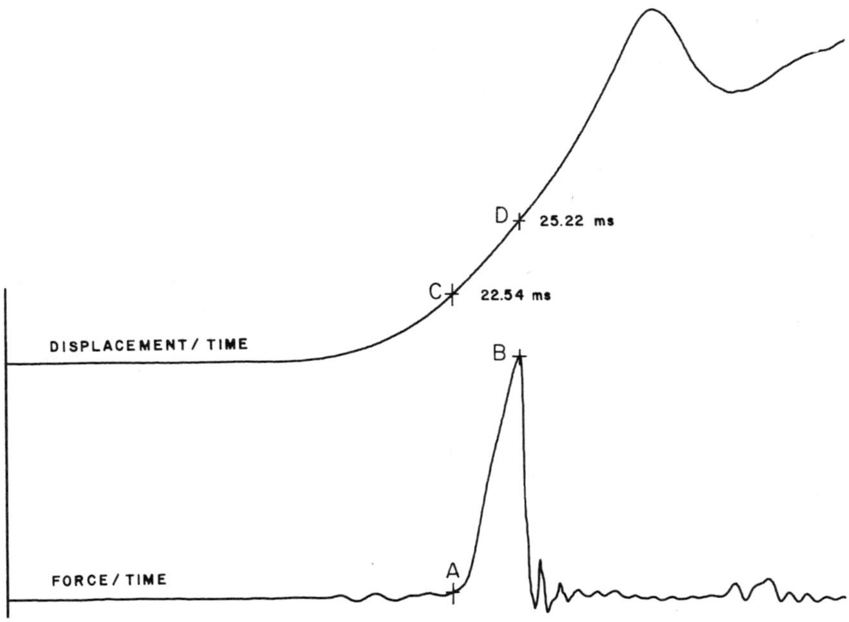

Figure 1. A typical output of the TIT; displacement/time, force/time.

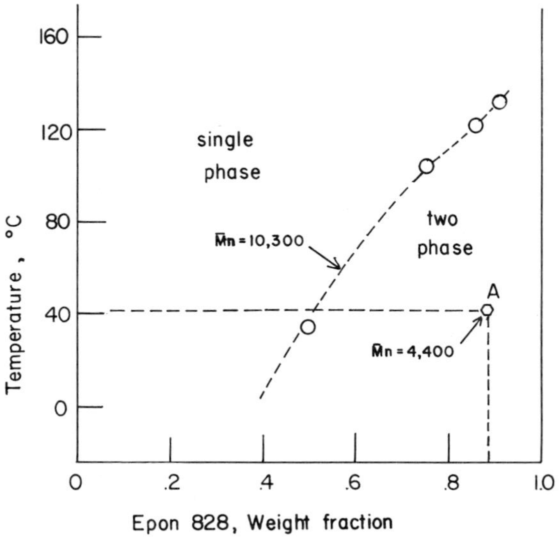

Figure 2. Typical Phase Diagram for a mixture of Epon 828 and PnBA rubber.

CTPnBA), at 40°C, was about $\bar{M}n$ = 4400. This is shown as point A in Figure 2. It was also observed that CTPnBA rubbers of higher molecular weights, up to $\bar{M}n$ = 6600, became miscible with Epon 828 at room temperature in the presence of selected curing agents; this study will be presented later.

The curing agent plays an important role in controlling the mechanical properties of epoxy resins. Most curing agents participate in the construction of the crosslinked epoxy network. Therefore, the nature of the curing agent and the epoxy resin/curing agent ratio have a major effect in shaping the properties and the morphology of the thermoset polymer. Although DMP-30 acts only as a catalyst in the curing process of Epon 828, the concentration of this tertiary amine has a substantial effect on the crosslink density and therefore on the impact strength of the crosslinked epoxy. CTPnBA/EPon 828 systems cured with various concentrations of DMP-30 were studied by notched Izod impact test, Table I. The CTPnBA used in this study had a molecular

TABLE I. Izod impacts strength of Epon 828/PnBA/DMP-30 systems and controls: Low functionacity RLP

#	Epon 828 parts	CTPnBA parts	DMP-30 parts	Impact strength Izod (lb-ft/in)
a-1	100	10	1.0	1.13±0.07
a-2	100	10	2.3	0.91±0.06
a-3	100	10	5.0	0.77±0.01
a-4	100	/	5.0	1.37±0.27
a-5	100	/	7.2	1.18±0.14
a-6	100	10	10.0	0.79±0.09
a-7	100	/	10.0	0.86±0.20

weight of $\bar{M}n$ = 3850, and a low functionality of f = 1.15-1.30. The resins were cured at 120°C/60 min and post cured at 150°C/180 min. The results indicate that the impact strength of the polymers containing the low functional CTPnBA rubber were in general lower than the controls, an observation which is in agreement with similar studies in the field (4,14). This behavior may be expected since the rubber is not bonded to the resin, and the rubber particles effectively act as "holes" in the matrix and reduce the strength.

The effect of the curing agent concentration on the morphology of the polymer-polymer composites was studied by SEM. Micrographs of the fractured surfaces of samples a-1, a-2, a-3 and a-4 are shown in Figures 3, 4, 5 and 6 respectively. The average density and diameter of the rubber particles are given in Table II. The number of particles per unit area increased with

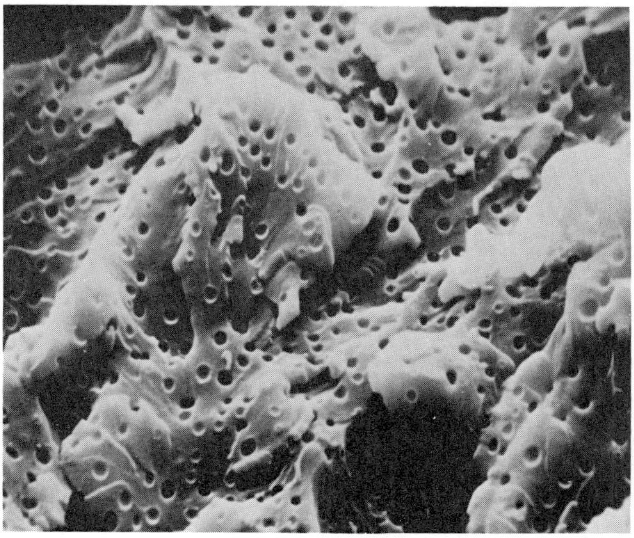

Figure 3. SEM micrograph of composite a-1, 1700X.

Figure 4. SEM micrograph of composite a-2, 1500X.

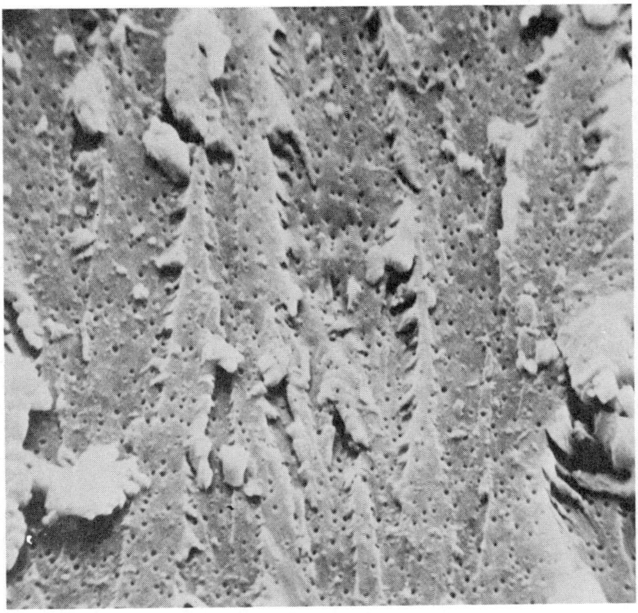

Figure 5. SEM micrograph of composite a-3, 1750X.

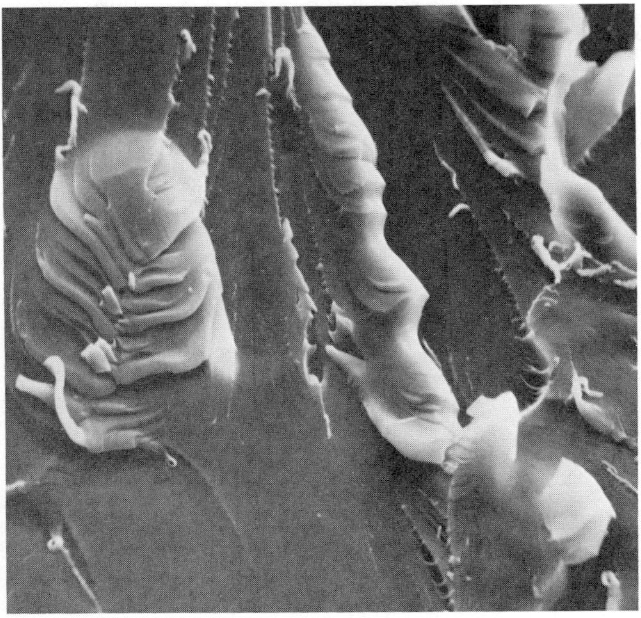

Figure 6. SEM micrograph of control a-4, 750X.

TABLE II. Average number and diameter of the rubber particles in Epon 828/CTPnBA/DMP-30 composites Table 1 gives the compositions.

#	# of particles per cm^2	average diameter (cm)
a-1	1.18×10^4	1.36×10^{-4}
a-2	1.48×10^4	9.18×10^{-5}
a-3	2.24×10^4	4.37×10^{-5}

increasing concentration of the curing agent, whereas the average diameter of the particles showed an opposite trend. The observations could be explained by the difference in the degree of solubility of the CTPnBA rubber in the epoxy resin, as affected by the various concentrations of the curing agent. This hypothesis is based on the enhanced miscibility of the rubber/epoxy system with the increasing concentration of DMP-30. Another hypothesis which might explain the observation is based on the difference in the precipitation rate of the rubber particles, during the curing process, at the various DMP-30 concentrations. This theory assumes that the crosslinking density of the epoxy resin is a function of the curing agent concentration.

From the data in Table I we see that the presence of a low functionality RLP in the epoxy network has reduced the impact strength of the crosslinked Epon 828. The effect of a higher rubber functionality on the impact strength of the epoxy/RLP system was studied with CTPnBA which was polymerized in bulk (21), and had a functionality of $f \gtrsim 1.8$. The composition of Epon 828/CTPnBA systems cured with three different curing agents are given in Table III. The CTPnBA rubber had a molecular weight

TABLE III. Composite and control compositions: Higher functionality RLP's.

#	Epon 828 + TBAI parts	CTPnBA parts	curing agent type	curing agent parts
3-a	100	9.9	DMHDA	27.0
3-b	100	25.0	DMHDA	27.0
3-c	100	9.9	DMP-30	1.4
3-d	100	10.0	DMP-30	4.9
3-e	100	10.0	MDA	37.9
3-f	100	/	DMHDA	27.7
3-g	100	/	DMP-30	5.0
3-h	100	/	MDA	39.2

average of $\bar{M}n$ = 6600, and a functionality of f = 1.91. The liquid rubber was initially mixed with the Epon 828, which contained 1% by weight of dissolved tetra-butylammonium iodide (TBAI). The TBAI catalyst was added to the mixture to enhance the reaction between the terminal carboxyl groups and the epoxide rings. The epoxy and the CTPnBA mixtures were not compatible at room temperature, but the addition of 2,5-dimethyl-2,5 hexanediamine (DMHDA) curing agent resulted in clearing of the epoxy/rubber solution; the "hazy" mixture became usually clear. The addition of the DMP-30 and the methylene dianiline (MDA) to the Epon 828/CTPnBA did not considerably change the miscibility of the mixture components. The systems were cured at 120°C/60 min followed by 150°C/180 min. Table IV summarizes the Izod and the tensile impact strength results of the polymer-polymer composites and the respective controls. Four significant results were observed.

TABLE IV. Impact strength of the composites and controls. ± Values represent standard deviations.

#	Izod	Tensile Impact Strength		
	Impact Strength (lb-ft/in)	σ_u (pa x 10^{-7})	ε_u (%)	Impact Strength (J/m^3 x 10^3)
3-a	1.43 ± 0.10	8.68 ± 0.77	15.7 ± 1.4	7.77 ± 1.57
3-b	1.51 ± 0.12	9.15 ± 0.63	18.6 ± 1.3	8.62 ± 1.02
3-c	0.89 ± 0.07	7.32 ± 0.15	12.9 ± 3.6	5.75 ± 1.09
3-d	1.01 ± 0.12	6.36 ± 1.33	10.5 ± 1.1	3.76 ± 0.75
3-e	1.13 ± 0.13	6.87 ± 1.36	15.8 ± 2.8	6.15 ± 0.65
3-f	1.40 ± 0.22	10.36 ± 0.63	9.7 ± 2.5	5.45 ± 2.55
3-g	1.73 ± 0.27	8.88 ± 0.49	9.2 ± 1.9	5.47 ± 1.48
3-h	1.68 ± 0.22	9.60 ± 0.32	9.1 ± 0.7	6.03 ± 0.31

1. The Izod impact strength of the composites remain in general lower than the controls.

2. The tensile impact energies of all composites (except for 3-d) are higher than the controls.

3. The high-speed ultimate strains of all the composites are larger than the control, up to 90% in the extreme case.

4. The ultimate stresses of the composites are slightly lower than the controls.

These facts, without further examining the details (type of rubber and curing agents) suggest that

a. Based on the Tensile Impact Tester results, the CTPnBA rubber has toughened the epoxy somewhat, primarily through increasing the ultimate tensile strain.

b. The toughening effect of the rubber was not evident from the Izod impact test. The toughening mechanism is thought to result from a microcavitation process. This phenomenon occurs throughout the gage length of the dogbone sample in the tensile tester, while the Izod fracture is limited to deformations in a short region of crack propagation in the sample. It appears that the rubber toughening mechanism cannot be fully appreciated by the nature of the fracture imposed on the notched sample in the Izod test. A typical stress-strain curve of a CTPnBA/Epon 828 composite is shown in Figure 7. The elastic tensile strain was followed by the "plastic" strain which was actually a deformation caused by a microcavitation mechanism throughout the whole gage length. The composites tested on the Tensile Impact Tester have "stress whitened" in the gage section, Figure 8. This whitening is an evidence for microcavitation in the composite, a phenomenon not observed in the control samples. Overall, the strain of the polymer-polymer composites appears to result from a combination of elastic deformation and microcavitation/plastic deformation. A "butterfly" pattern is observed upon bending or tensile deformation of the composites (fig. 9). The stress whitening and 'butterfly' pattern both disappear upon mild heating; the sample after heating appears the same as before deformation. Some composite samples (cured with DMHDA) have also experienced shear yielding during the tensile impact test, in addition to the microcavitation deformation, Figure 10. Shear yielding was observed, although to a lesser degree, with the controls of the DGEBA/DMHDA systems.

The effect of the curing cycle on the impact strength of the CTPnBA/Epon 828 system was found to be significant. The effect was demonstrated in a two step curing cycle procedure. In the first step of the procedure 40 parts of CTPnBA rubber ($\bar{M}n \sim 6600$, $f = 1.91$) were mixed with 60 parts of Epon 828, catalyzed by 1% TBAI. The mixture was reacted at 120°C for 2 hrs. During this stage the carboxyl-terminated n-butyl acrylate reacted with the Epon 828 to near completion; titration of the mixture after the 2 hrs. reaction showed essentially no carboxyl groups remaining in the solution. The epoxy resin, in large excess, had reacted with the CTPnBA to form an epoxy/rubber/epoxy intermediate compound. In the second step, the intermediate compound was dissolved in more epoxy to form a mixture of 10 parts CTPnBA in 100 parts Epon 828. Then, the curing agent (DMHDA) was added and the mixture was cured at 120°C/60 min and then 150°C/120 min. The two step curing cycle resulted in a tougher composite having an Izod impact strength of 1.70±0.20 ft-lb/in of notch, an improvement of 21% over the control sample 3f, and a tensile impact strength of $(10.80 \pm 2.2) \times 10^{-3}$ (J/m^3), an improvement of 98% over the control. These results suggest the two step curing cycle was better than the single step procedure. Samples made by the two step curing cycle procedure showed appreciable stress whitening in the tensile impact test.

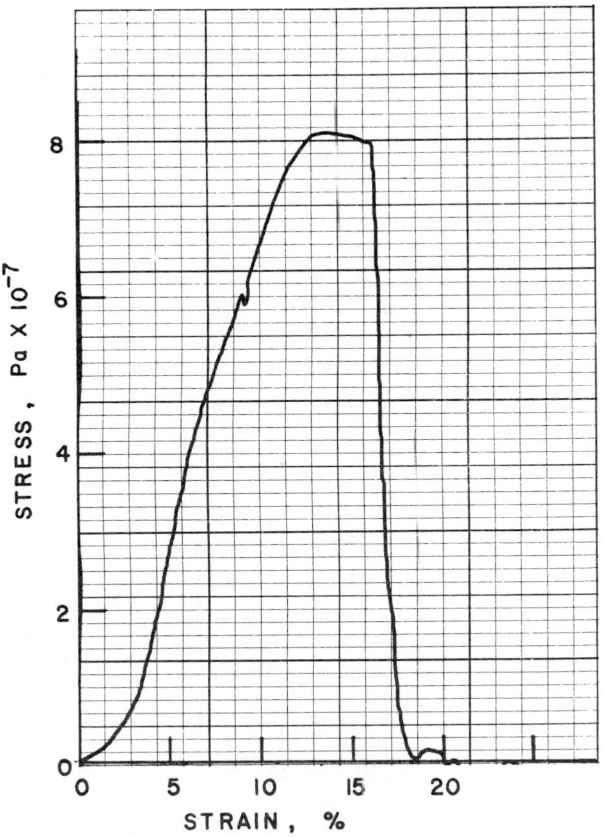

Figure 7. A typical stress-strain diagram of composite 39-b, measured by the TIT technique.

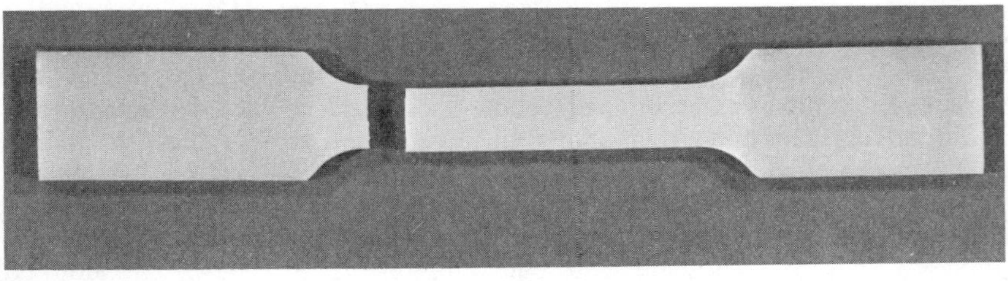

Figure 8. Stress whitening of composite 39-b formed by tensile stress in the tensile impact test.

Figure 9. Microcavitation in a "butterfly" pattern formed in Epon 828/CTPnBA cured by DMHDA in the two step procedure.

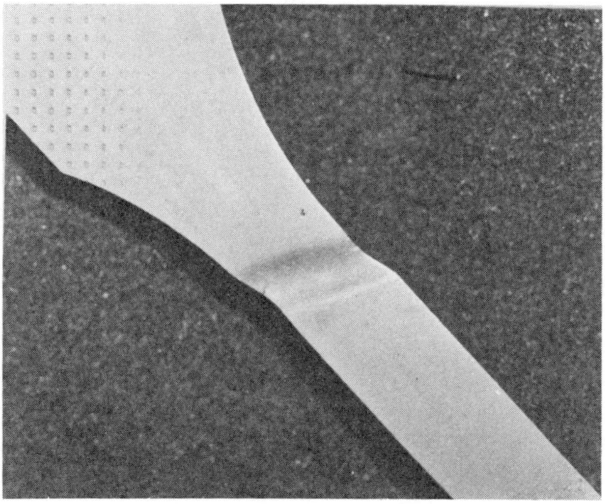

Figure 10. Shear band formation in a tensile test of the DGEBA/DMHDA system.

The glass transition temperature of the PnBA composite cured using the two step procedure was only about 3°C lower than the 3f control without rubber, Figure 11. This shows that the mechanism of rubber toughening was not simple plasticization. Interestingly, the β-transition of the composite appeared to be slightly higher than the Epon 828-(DMHDA) control.

Conclusions

The potential of CTPnBA as an RLP for epoxy resins was shown to be promising. For effective toughening, the poly n-butyl acrylate must bond to the matrix. Carboxyl terminal groups on PnBA were found efficient for this purpose. The type of curing agent and the curing procedure were found to be primary factors in controlling the toughening of the composite; DMHDA was found most effective curing agent for the DGEBA/CTPnBA system. The two step curing process gave best impact strength. It was obsrved that the toughening mechanism was mainly by a microcavitation process, however, some shear deformation was evident in a few of the tested samples.

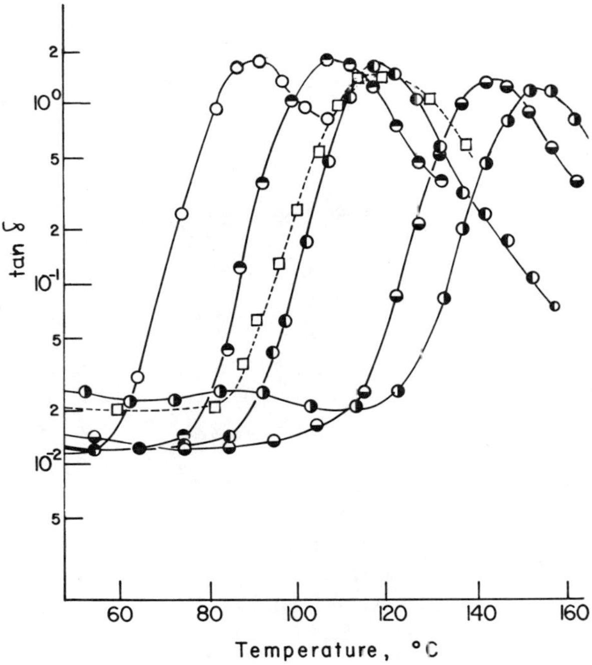

Figure 11. The mechanical damping of Epon 828/DMHDA system vs. temperature at various amine excess formulations, ◑ - 0%, ◒ - 25%, ◐ - 50%, ◓ - 75%, O - 100%, □ - CTPnBA modified system, cured in the two step procedure.

Acknowledgment

The authors express their appreciation to Shell Development Company for support of this research.

Literature Cited

1. Lee, H.; Neville, K. "Handbook of Epoxy resins," McGraw-Hill, New York, 1967.
2. Manson, J. A.; Sperling, L. H. "Polymer Blends and Composites," Plenum Press, New York, 1976.
3. Perez, R. J. in "Epoxy Resin Technology," P. S. Bruins, Ed., Interscience, New York, 1968, p. 45.
4. McGarry, F. J.; Willner, A. M. "Toughening of Epoxy Resin by an Elastomeric Second Phase," R68-8, March, MIT, 1968.
5. Manzione, L. T.; Gillham, J. K. American Chemical Society Organic Coatings and Plastics Chemistry Preprints, 1979, 41, 364.
6. Manzione, L. T.; Gillham, J. K. American Chemical Society Organic Coatings and Plastics Chemistry Preprints, 1979, 41, 371.
7. Bucknall, C. B.; Yoshii, T. The British Poly J., 1978, 10, 53.
8. Manzione, L. T.; Gillham, J. K.; McPherson, C. A. J. Appl. Polym. Sci. 1981, 26, 889.
9. Manzione, L. T.; Gillham, J. K.; McPherson, C. A. J. Appl. Polym. Sci. 1981, 26, 907.
10. Meeks, A. C.; Polymer, 1974, 15, 675.
11. Walker, J. W.; Richardson, W. E.; Smith, C. H. American Chemical Society Organic and Plastics Chemistry Preprints, 1975, 35, 333.
12. Noshay, A.; Robeson, L. M. J. Polym. Sci., Chem. Ed., 1974, 12, 689.
13. Sperling, L. H.; Friedman, D. W. J. Polym. Sci., part A-2, 1969, 7, 425.
14. Scarito, P. R.; Sperling, L. H. Poly. Eng. and Sci., 1979, 19, 297.
15. Rowe, E. H. 26th Annu. Tech. Conf., Reinforced Plastics/Composites Div., SPI, Sec. 12-E, 1, 1971.
16. Riew, C. K.; Rowe, E. H.; Siebert, A. R. in "Toughness and Brittleness of Plastics", ADVANCES IN CHEM. SERIES, No. 154, American Chemical Society, Washington, D. C., 1976, p. 326.
17. Rowe, E. H. 24th Annu. Tech. Conf. Reinforced Plastics/Composites Div., SPI, Sec. 11-A, 1, 1969.
18. Rowe, E. H.; Siebert, A. R.; Drake, R. S. Modern Plastics, 1970, 47, 110.
19. Drake, R.; Siebert, A. SAMPE Quarterly, July, 1975.
20. Gazit, S. "Toughening of Epoxy Resin by Acrylic Elastomer," Ph.D. Dissertation, University of Connecticut, Sept. 1980.
21. Gazit, S.; Bell, J. P. (another paper in this volume)

RECEIVED December 16, 1982

5

Synthesis, Morphology, and Thermal Stability of Elastomer-Modified Epoxy Resins

WILLIAM A. ROMANCHICK, JOHN E. SOHN, and JON F. GEIBEL

Western Electric Company, Inc., Engineering Research Center,
Princeton, NJ 08540

The incorporation of elastomers into epoxy resins has been an active area of research over the past decade (**1-11**). The primary reason for this interest has been the improved toughness of the modified materials. We have limited our work to the carboxy-terminated butadiene-acrylonitrile copolymers (CTBN) produced by B.F.Goodrich. These materials have been reacted with liquid epoxy resins and curing agents previously by several authors. Riew, Rowe, and Siebert have been particularly active in this area studying such things as the effect of rubber domain size on maximum toughness (**6**). The materials described by these authors were not applicable to our area of interest, functional powder coatings. We have succeeded in synthesizing solid rubber-modified epoxy resins with epoxide equivalent weights (EEW) ranging from 800-2000 g eq^{-1}. The resins are capable of being dry blended with other resins, curing agent(s) and additives, then extruded and milled to give superior coating powders. These powders coated either by electrostatic spray or fluid bed, give exceptional adhesion to the substrate when compared to non-modified epoxies. This paper will concern itself with the synthesis and mechanism in the manufacture of solid rubber-modified resins, the morphology and thermal stability of these resins, and some of the properties of cured coatings made with these materials.

Synthesis

Solid epoxy resins can be made from the diglycidyl ether of bisphenol A (BPA) via a chain extension reaction with bisphenol A. Keeping bisphenol A as the limiting reagent assures the product will still be a bisepoxide. One equivalent of phenol reacts with one equivalent of epoxide as shown in Reaction Scheme 1.

The reaction in our laboratory is catalyzed by triphenylphosphine. The reason for using this catalyst and a proposed mechanism for the reaction will be discussed later. The mathematics used to calculate a desired chain length of the solid bisepoxide are straightforward. The number of equivalents of epoxide in the product must equal the number of equivalents of epoxide in the reactant minus the number consumed during the advancement reaction. Letting X be defined as the weight percent diglycidyl ether of bisphenol A (titrated epoxy equivalent weight = 190 g eq^{-1}) and Y being defined as the weight percent of bisphenol A (equivalent weight = 114 g eq^{-1}), one arrives at a general expression for conservation of epoxides for advancement to any desired EEW.

$$\frac{100}{EEW} = \frac{X}{190} - \frac{Y}{114} \qquad (1)$$

0097–6156/83/0221–0085$09.50/0

Conservation of mass dictates that the sum of the weights of the diglycidyl ether of bisphenol A and bisphenol A must equal the weight of the product.

$$X + Y = 100 \qquad (2)$$

Thus, with two equations and two unknowns, a unique solution to X and Y can be calculated. Carboxy-terminated copolymers of butadiene and acrylonitrile (CTBN) can be incorporated into the epoxy resin through a simple esterification reaction. One equivalent of carboxylic acid is esterified by one equivalent of epoxide. This reaction is again catalyzed by triphenylphosphine. Since, for our reactions, the limiting reagent in the esterification is the CTBN elastomer, the product is an elastomer capped with an epoxy resin. The epoxy terminated rubber is then capable of reacting in the usual manner with curing agents or in the previously discussed advancement reaction (Reaction Scheme 2).

The amount of elastomer contained in a rubber-modified epoxy resin is usually dictated by the final application and material properties desired. Knowing that one equivalent of epoxide reacts with one equivalent of carboxylic acid and the final concentration of elastomer that is desired, one can add another term to Equation (2) to account for the equivalents of epoxide consumed during the esterification reaction. The equivalent weight of the CTBN elastomers vary from lot to lot but is typically $\sim$1800 g eq^{-1}. Using this number as an example, the expanded equation now reads,

$$\frac{100}{EEW} = \frac{X}{190} - \frac{Y}{114} - \frac{Z}{1800} \qquad (3)$$

where Z is the weight percentage of CTBN elastomer. Conservation of mass must now include the third component.

$$100 = X + Y + Z \qquad (4)$$

Thus, since Z is chosen by the formulator, we have two equations and with two unknowns. A unique solution for X and Y can be calculated.

Performing both advancement and esterification reactions results in a solid rubber-modified epoxy resin. The advancement and esterification reactions may be performed in several ways: concomitantly, esterification followed by advancement, or advancement followed by esterification.

Proposed Mechanisms

The basic catalysis of the epoxide-phenol reaction has been studied by a variety of authors (**12,13**). In general, the conclusion has been that basic catalysis gives selectivity for the phenol-epoxide reaction over the secondary alcohol-epoxide reaction (the secondary alcohol being formed from the opening of an epoxide). Although the base catalyzed reaction has been studied, the specific mechanism for triphenylphosphine catalysis has not been determined. We propose the mechanism shown in Reaction Scheme 3 which is consistent with our experimental results.

When all the bisphenol A has reacted, a plausible mechanism accounting for the destruction of the catalyst is shown in Reaction Scheme 4 (Note: triphenylphosphine can also be air oxidized at reaction conditions to triphenylphosphine oxide).

Nucleophilic attack by the triphenylphosphine opens the epoxide, producing a betaine, 1. Proton abstraction from bisphenol A yields the phenoxide anion, 2. The phenoxide reacts with the electrophilic carbon attached to the positive phosphorus, 3, regenerating the catalyst. When the phenol is exhausted, the betaine can decompose into a terminal olefin and triphenylphosphine oxide (the final step in the Wittig

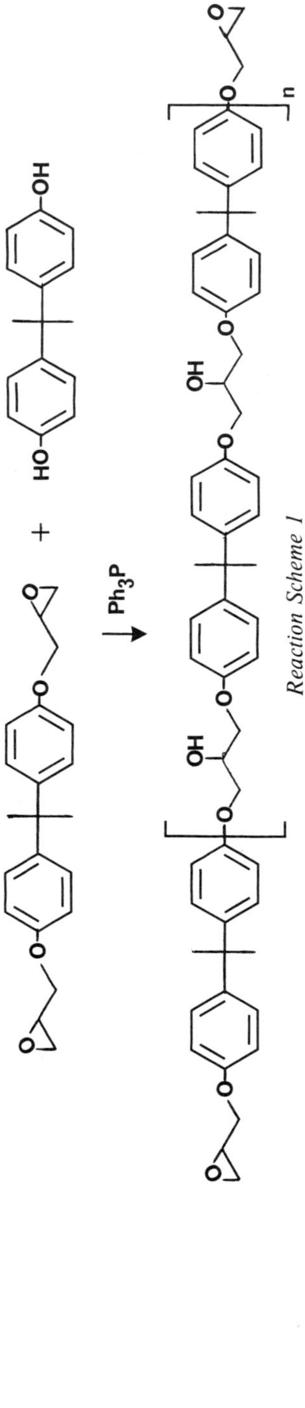

Reaction Scheme 1

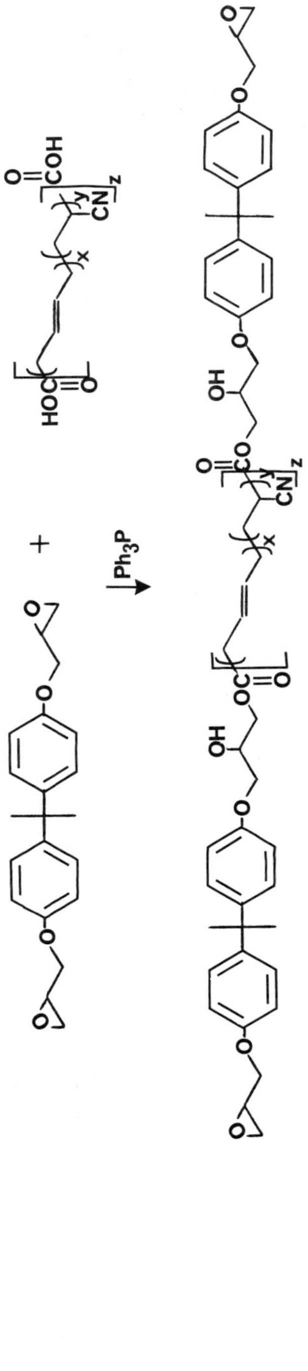

Reaction Scheme 2

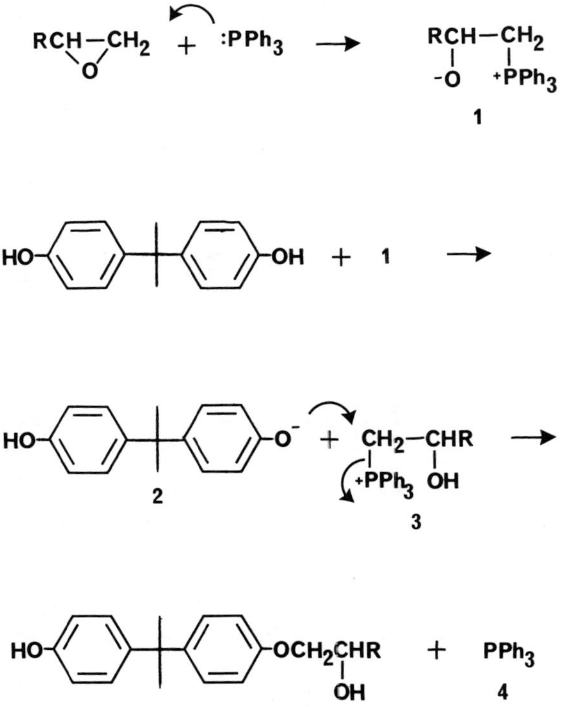

Reaction Scheme 3

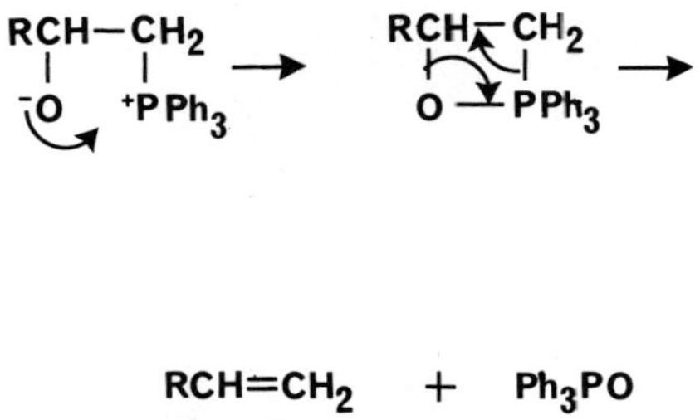

Reaction Scheme 4

reaction). Triphenylphosphine oxide is an unreactive material whose presence does not degrade the properties of the resin. Triphenylphosphine is used in catalytic quantities, therefore insignificant amounts of terminal olefin are generated, which is not reactive toward curing agents.

The role of the triphenylphosphine is similar in the rubber-modification reaction. In this case however the betaine abstracts a proton from the carboxylic acid. The carboxylate anion attacks the electrophilic carbon attached to the phosphorus forming an ester linkage and regenerating the catalyst. The rubber is terminated on both ends by epoxides and can later be crosslinked into the epoxy matrix during the coating and curing phases. This sequence is shown in Reaction Scheme 5.

Morphology

The variables previously studied which impact on morphology include elastomer type and concentration, curing agent type and concentration, cure conditions, and time to gelation. It has been concluded that the morphology is determined during the curing process, with phase separation of the elastomer from the solution, resulting in the formation of segregated domains of elastomer particles covalently bound to the epoxy matrix. These results were obtained from liquid systems, where the rubber was not prereacted with epoxy. We have investigated the morphology of solid systems, where the elastomer has been prereacted with epoxy. The molecular weight distribution in our system is such that phase separation is evident even in the solid uncured system. The analytical techniques used to determine the morphology include scanning electron microscopy (SEM), transmission electron microscopy (TEM), torsional braid analysis (TBA), and differential scanning calorimetry (DSC). The factors which contribute to morphological definition which we studied include rubber type and concentration, curing agent type and concentration, and cure conditions.

The solid rubber-modified epoxy was prepared from bisphenol A advanced EPON Resin 828, the CTBN elastomer, and either an EPON P100 series curing agent (Shell) or 2-methylimidazole. The resin was prepared as discussed above; the curing agent was added to the modified resin by melt mixing, and the resulting material pulverized prior to the curing process. Three carboxy-terminated elastomers were used in this work; they are listed in Table I along with their glass transition temperatures (T_g), acrylonitrile content, and solubility parameters.

Table I

Elastomer Properties

Elastomer	T_g(C)	% Acrylonitrile	Solubility Parameter
CTB 2000X162	-74	0	8.04
CTBN 1300X8	-45	18	8.77
CTBN 1300X13	-30	26	9.14

Phase separation is present in the uncured resin prior to incorporation of the curing agent (**11**). The two phases occur as core-shell structures, with the interior consisting of epoxy, the shell, the phase separated elastomer, and the matrix is epoxy. Figure 1 is a TEM photomicrograph showing the core-shell morphology of a 15 wt% 1300X8 modified epoxy; the rubber portion was defined by osmium tetroxide staining.

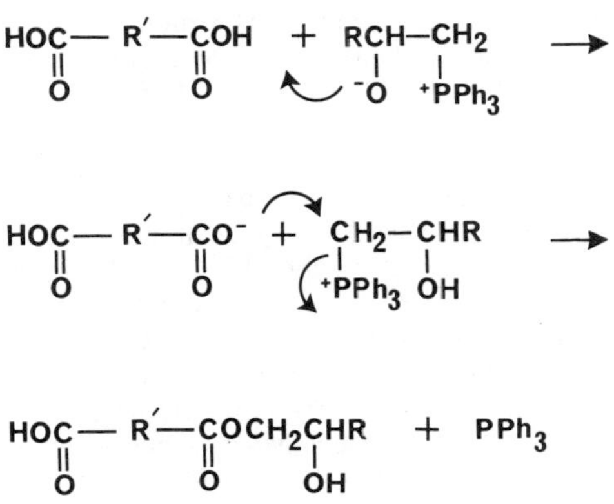

Reaction Scheme 5

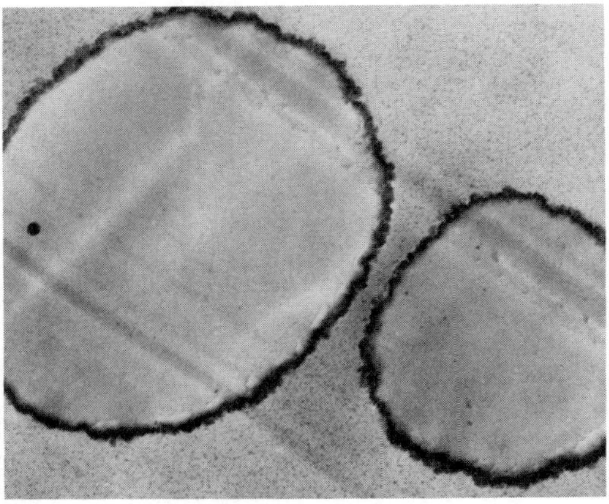

Figure 1. TEM micrograph of 15% X8 at 20,000X.

Tetrabromobisphenol A was incorporated into one resin system at a concentration to yield 1 atom% bromine. Comparison of the bromine concentration of the interior and exterior of the core-shell structures, obtained by energy dispersive x-ray analysis (EDAX), demonstrates that the interior of the particle has the same bromine concentration as the matrix.

The average particle diameter for the 15 wt% 1300X8 system is 2.78μm. The 15 wt% 1300X13 system possesses a much different distribution of particles - very small (~ 300Å diameter) rubber containing domains are present, and the diameters of the larger domains fall into two regions, 0.30μm and 1.40μm. These results are explainable by the difference in rubber-epoxy compatibility in the two systems. Both the epoxy and rubber glass transition temperatures were determined for the two systems by TBA, the values are given in Table II along with the values for the corresponding pure epoxy and pure rubber.

Table II

Rubber - Epoxy Compatibility

	Rubber T_g(C)	Epoxy T_g(C)
Pure Epoxy	-	63
Pure 1300X8	-45	-
Pure 1300X13	-30	-
15 wt% 1300X8	-45	58
15 wt% 1300X13	-5	55

The 1300X13 system is more compatible than the 1300X8 system as evidenced by the significant increase in the rubber T_g of the 1300X13 system from the T_g of the pure 1300X13 elastomer; no significant increase in the rubber T_g of the 1300X8 system is observed. The greater compatibility of the 1300X13 elastomer with the epoxy is expected from the greater acrylonitrile content of the 1300X13 rubber, and the higher solubility parameter. The greater 1300X13-epoxy compatibility allows the small particles to exist since there is much less of a driving force to reduce rubber-epoxy interaction by forming larger particles with significantly less rubber surface area.

The morphology of the uncured material prior to cure, but after incorporation of the curing agent was determined by TEM. The morphology was altered by the melt mix process; much smaller particles were observed with average diameters of 0.11μm, with no evidence of core-shell structures.

The effect of elastomer composition and concentration on morphology was determined from a series of resins cured with EPON CURING AGENT P-108, a substituted imidazole accelerated dicyandiamide. The morphology was determined by TEM, and the glass transition temperatures were determined by TBA and DSC. The resins were cured for 0.5h at 200C; and gelation is attained in approximately one minute. The morphological data are presented in Tables III, IV, and V.

Table III

Morphology of 2000X162 Modified Epoxy

wt% 2000X162	Av. Part. Diam (μm)	Range (μm)
7.5	0.25	0.05-1.00
10.0	0.35	0.05-1.00

Table IV

Morphology of 1300X8 Modified Epoxy

wt% 1300X8	Av. Diam. (μm)	Range (μm)	ET_g(C)	RT_g(C)
2.5	0.16	0.02-0.30		
5.0	0.23	0.02-0.54	109	-67
7.5	0.22	0.04-0.54	106	-56
10.0	0.30	0.04-0.68		
12.5	0.37	0.04-1.00		
15.0	0.48	0.12-1.16	102	-53

Table V

Morphology of 1300X13 Modified Epoxy

wt% 1300X13	Av. Diam. (μm)	Range (μm)	ET_g(C)	RT_g(C)
2.5	0.14	0.06-0.34		
5.0	0.18	0.06-0.30	105	-39
7.5	0.25	0.04-0.52	103	-36
10.0	0.29*	0.04-0.68	100	-38
12.5	0.36*	0.12-0.76		
15.0	0.40*	0.12-0.76	97	-31

* A second distribution of very small (200-500Å) rubber domains is present, but not included in the values.

The rubber domains are found to exist as core-shell structures at all concentration levels of all three elastomers (Figure 2).

Several trends are evident from the data. A decrease in acrylonitrile content, with a corresponding reduction in epoxy-rubber compatibility, results in larger rubber domains. Reduced rubber-epoxy compatibility yields a greater degree of phase separation, which is apparent from comparison of the glass transition temperatures in the 1300X8 and 1300X13 systems. The T_g of the epoxy phase occurs at a lower temperature in the 1300X13 material as compared to the 1300X8 system; the epoxy T_g in the corresponding unmodified system is 112C. This is attributed to a greater degree of plasticization of the epoxy phase by the rubber in the 1300X13 modified system.

An increase in elastomer concentration results in larger rubber domains, a decrease in the epoxy T_g and an increase in the rubber T_g in both the 1300X8 and 1300X13 systems. The changes in the glass transition temperatures are due to increased plasticization of the epoxy phase with increased elastomer concentration. The

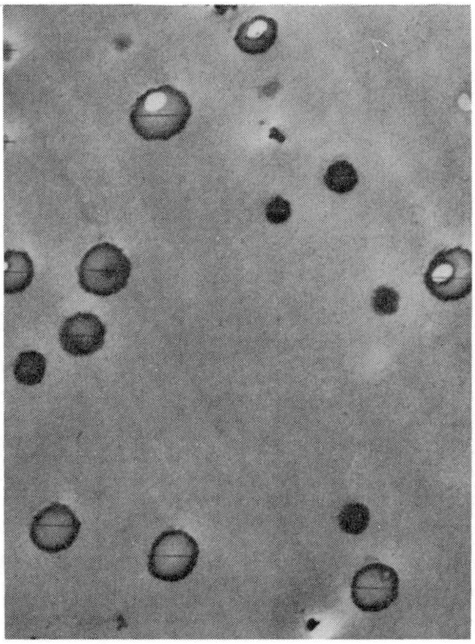

7.5% X8

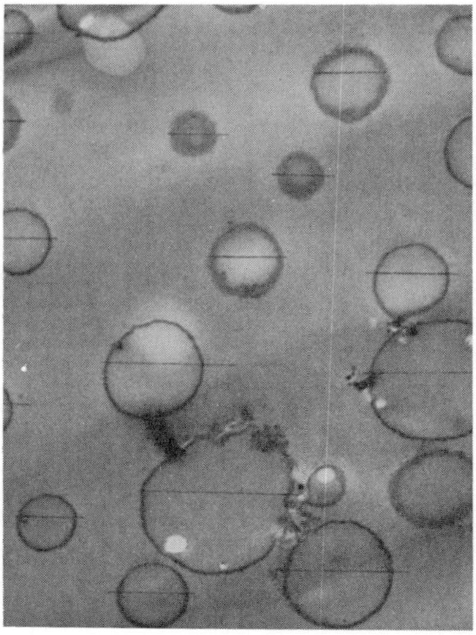

15% X8

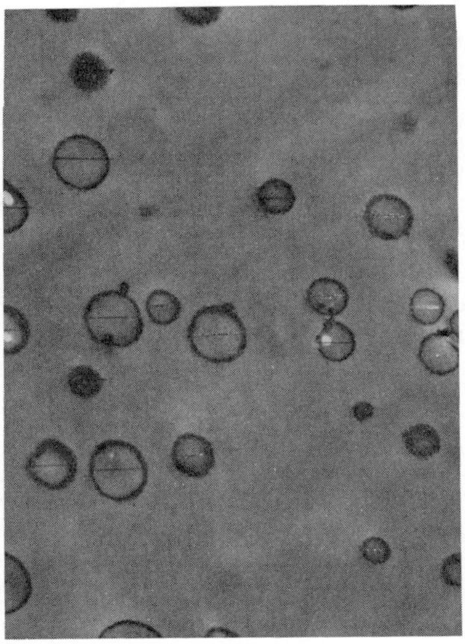

7.5% X13

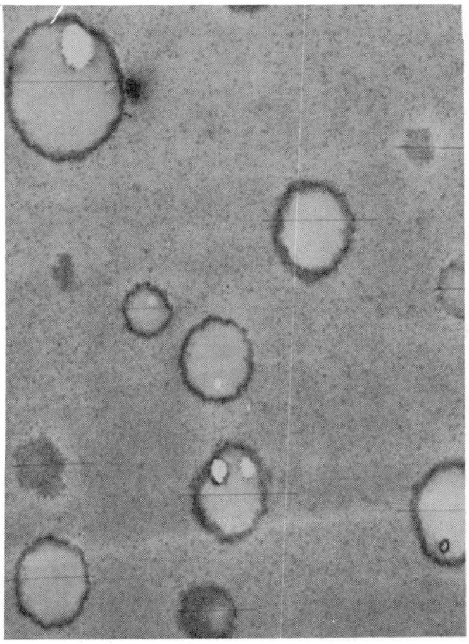

15% X13

Figure 2. Morphology of Cured Elastomer Modified Epoxies.

rubber T_g occurs at a lower temperature compared to the T_g of the pure elastomer. Others have observed this phenomenon and attribute the depression to differences in the coefficients of thermal expansion between the glassy epoxy matrix and the rubber phase (**9**). The larger coefficient of thermal expansion of the rubber phase results in constraint of the rubber domains upon cooling below the glass transition temperature of the matrix.

Further evidence of increased compatibility with increased acrylonitrile content is seen in the presence of small (200-500Å) rubber domains in the 1300X13 modified system at concentrations of 10, 12.5, and 15 wt%. The existence of these domains is a consequence of increased compatibility in the 1300X13 system; otherwise the rubber would tend to exist in larger core-shell structures where epoxy-rubber interactions are minimized.

The free energy of domain formation for the three elastomers in epoxy was calculated from the interfacial tension as determined by the pendant drop technique (**14**). The calculated free energies are listed in Table VI. The free energy becomes more negative with increasing acrylonitrile content. The energy is dissipated by increasing the surface area to volume ratio of the elastomer; the greater the energy to be dissipated the greater the driving force to form domains. Thus the large negative free energy of domain formation for 1300X13 results in smaller rubber domains as compared to the other rubbers and can explain the presence of the 200-500Å particles observed in the 1300X13 modified epoxies.

Table VI

Free Energy of Domain Formation

Elastomer	% Acrylonitrile	ΔG_d (J cm^{-3})
2000X162	0	-2
1300X8	18	-84
1300X13	26	-764

Effect of Curing Agent Type and Composition The glass transition temperatures, time to gelation at 200C, and morphologies were determined for 7.5 wt% 1300X13 modified epoxy cured with different curing agents. The curing agents are composed of different amounts of dicyandiamide and substituted imidazole. The effect of curing agent concentration was studied by varying the amount present in one series. All materials were cured for 0.5h at 200C; the data are presented in Table VII.

The EPON P series curing agents are composed of varying amounts of dicyandiamide and substituted imidazole accelerator; P108 contains the least amount of accelerator, P100 is composed purely of accelerator, and P104 is intermediate in composition between P108 and P100. Larger domains result as gel time increases in the P108 series, consistent with the theory that gelation determines morphology (**7,9**). This relationship holds for the other curing agents as well, with the smallest domains found in the system with the shortest time to gelation (2-methylimidazole cured). These data indicate that the morphology is determined by the gelation time.

Table VII

Morphology of 7.5 wt% 1300X13 Modified Epoxy vs Curing Agent

Curing Agent	P108	P108	P108
phr	3.3	2.8	2.3
RT_g(C)	-35	-38	-39
ET_g(C)	104	102	101
Gel Time (s)	50	78	105
Av. Diam. (μm)	0.20	0.28	0.27
Range (μm)	0.06-0.36	0.11-0.47	0.11-0.51

Curing Agent	P104	P100	Methylimidazole
phr	2.0	1.0	1.0
RT_g(C)	-35	*	-30
ET_g(C)	99	82,93**	104
Gel Time (s)	30	32	12
Av. Diam. (μm)	0.19	0.15	0.09
Range (μm)	0.06-0.29	0.06-0.23	0.04-0.15

* No RT_g was observed.
** Second value obtained after heating at 200C for 2h.

Effect of Cure Time and Temperature The impact of cure time (degree of cure) on morphology was determined using a 7.5 wt% 1300X13 modified epoxy cured with 70 wt% of the manufacturer's recommended level of P108. Gelation occurs after approximately one minute at 200C; solvent extraction yields no soluble material. Full cure is attained after 0.5h at 200C as determined by solvent absorption and torsion pendulum analysis. The morphological results are listed in Table VIII and illustrated in Figure 3.

Table VIII

7.5 wt% 1300X13 Modified Epoxy EPON® P108 Cured at 200C

Sample	Cure Time(min)	Av. Part. Diam.(μm)	Range(μm)
1	1	0.13*	0.06-0.25
2	2	0.10	0.05-0.16
3	7.5	0.21	0.10-0.34
4	30	0.20	0.08-0.32

* Many very small (200Å) particles present - not used in diameter or range calculations. Particles were not observed in other samples.

A bimodal particle distribution is observed in sample 1, which has just reached gelation. The morphology continues to change after gelation; after two minutes the very small particles (~200Å) are no longer present and the average particle diameter is only 50% that of the fully cured material. A static condition is reached by 7.5 minutes,

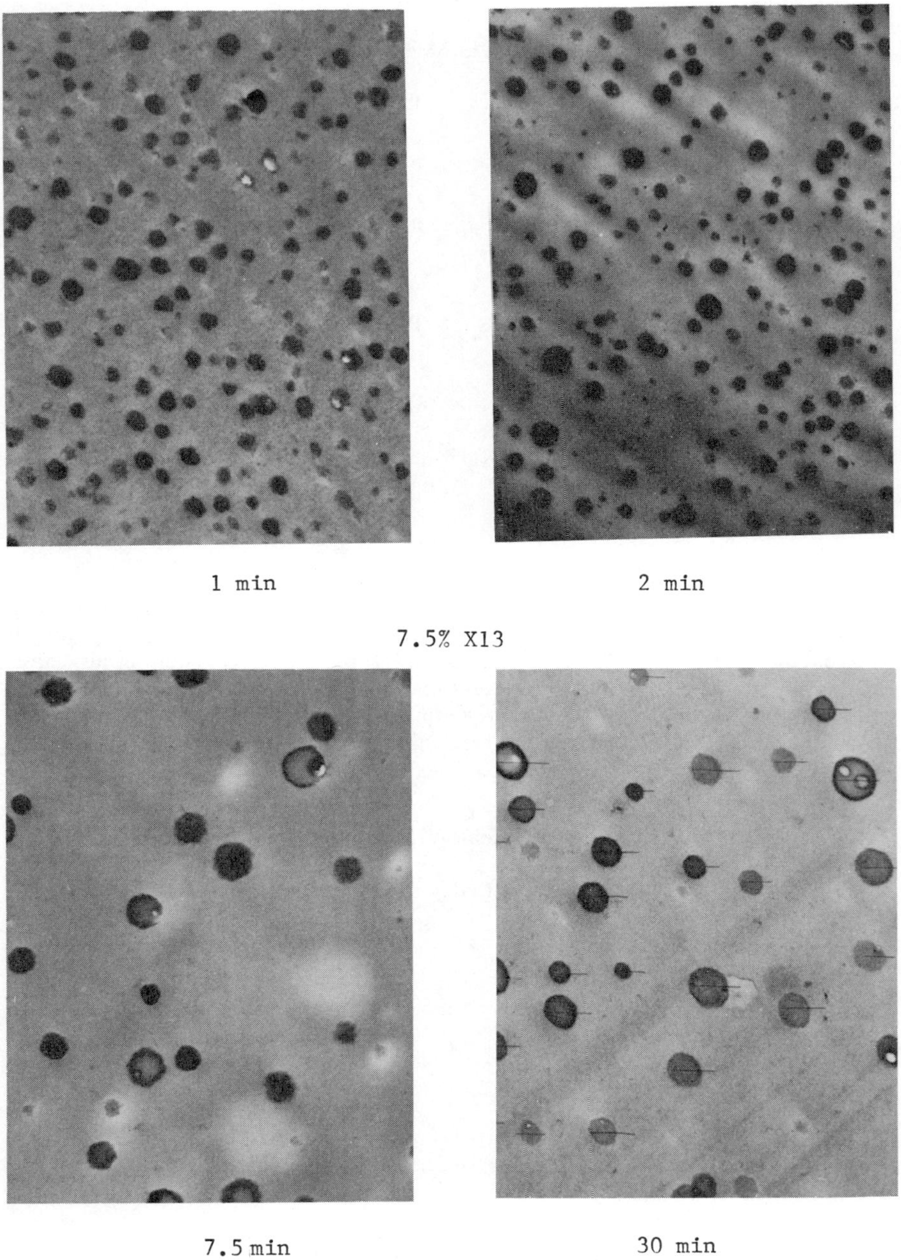

Figure 3. Effect of Cure Time on Morphology.

with the morphology nearly identical to that of the fully cured system. These results contrast with those obtained by TBA and cloud point determination which indicate that the morphology is frozen by gelation (**7,9**). These techniques, however, are not capable of elucidating the actual morphology of the system. The TBA spectra have been obtained for the uncured resin, and after curing at 200C for 2, 5, and 30 minutes (Figure 4). There are minimal differences between the spectra of the samples cured for 2, 5, and 30 minutes - the rubber glass transition temperatures are nearly identical, the epoxy glass transition temperature increases with cure time as the system becomes fully cured.

Identical morphologies were obtained from two samples of the same resin system with significantly different cure conditions - sample 4: 200C/0.5h, sample 5: 170C/4h (Table IX). Solvent absorption and torsion pendulum results were used to determine that sample 5 did not achieve full cure until 4 hours. The times to gelation differ by a factor of three, yet the final morphologies are nearly identical.

Table IX

7.5 wt% 1300X13 Modified Epoxy Cured with EPON® P108

	Sample 4	Sample 5
Cure Temperature (C)	200	170
Cure Time (h)	0.5	4
Gel Time (s)	59	155
Av. Part. Diam. (μm)	0.20	0.20
Range (μm)	0.07-0.35	0.08-0.32

Conclusion The morphology of solid elastomer-modified epoxy resins has been determined using TEM, TBA, and DSC. The data show that the epoxy and elastomer are incompatible in both the uncured and cured states, with phase separation evident and the rubber existing in core-shell structures. The morphology is dependent upon elastomer composition and concentration, curing agent composition and concentration, and cure conditions. Morphological changes are found to occur after gelation in these systems, with the final morphology dependent on the parameters previously listed.

Thermal Stability

The objective of this section is to characterize the thermal stability of uncured solid rubber-modified epoxy resins. The effects of extended thermal history on melt viscosity and epoxide equivalent weight are discussed. The influence of the type and concentration of CTBN elastomer in the solid rubber-modified epoxy resin on melt viscosity and EEW is also discussed. Mechanistic considerations are proposed to explain the side reactions which influence the thermal stability of solid rubber-modified epoxy resins.

Results and Discussion To contrast the thermal stability of rubber-modified epoxy resins, a control experiment was performed to establish the thermal stability of a non-rubber-modified epoxy resin. An advancement reaction was performed (EEW 900 g eq^{-1}), the product maintained isothermally at 175C, and the reaction progress followed by thermally quenching aliquots to room temperature at various time intervals. Although previously published reports indicate no significant increase in the EEW of an

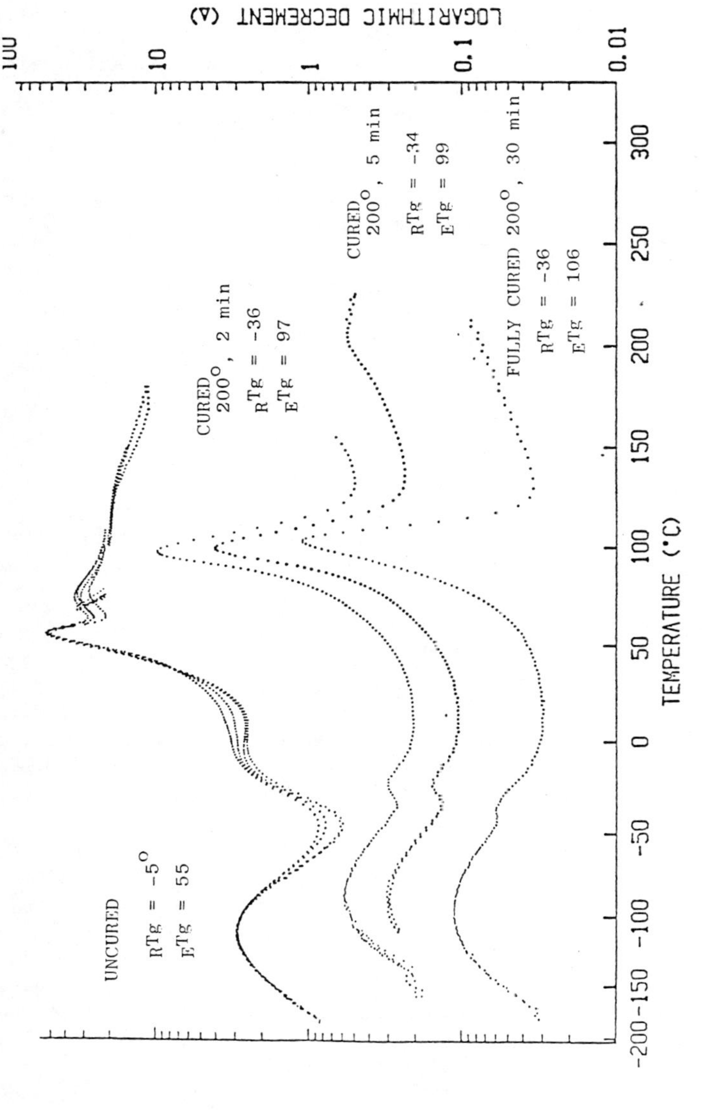

Figure 4. TBA Spectra of 1300X13 Modified Epoxy.

epoxy resin owing to isothermal aging of the resin for several hours at ~ 175C (**15**), our data indicate a monotonic upward trend for the EEW. The resin was held at 175C with efficient stirring and no special precautions were taken to exclude air. Figure 5 shows the time dependent behavior for the change of the EEW during isothermal aging for this non-modified epoxy resin. The EEW increases at a rate of 8.8g $eq^{-1}h^{-1}$. The aging process is thus characterized by a loss of epoxides with time.

The loss of epoxides is accompanied by an increase in melt viscosity. Figure 6 shows the increase in melt viscosity measured at 175C versus time of isothermal aging at 175C for this non-modified epoxy resin. The melt viscosity increases dramatically by more than a factor of two during seven hours of isothermal aging. The side reaction that consumes epoxides occurs concommitantly with the growth of larger molecules as supported by the increased melt viscosity. The melt viscosity increases at an accelerating rate throughout the isothermal aging.

The effect of rubber modification on the rate of disappearance of epoxides and on the rate of increase of melt viscosity is of interest because it may influence the end use formulation stoichiometry (i.e., amount of curing agent). A rubber-modified epoxy resin (theoretical EEW = 900 g eq^{-1}, 10 wt% 1300X13) was isothermally aged at 175C in the same manner described above. Figure 7 shows the EEW of aliquots drawn at various time intervals during the aging experiment. The same linear time dependence for the increase of EEW is observed, however, the rate of reaction is increased by 65% to 14.5 g eq^{-1} h^{-1}. Figure 8 shows the rapidly increasing melt viscosity during isothermal aging of this rubber-modified epoxy resin. That the rate of loss of epoxides is greater for the rubber-modified epoxy resin than for the analogous non-modified epoxy resin implies a lower inherent thermal stability for the uncured solid rubber-modified epoxy resin. The thermally induced side reaction which results in the loss of epoxides appears to be either enhanced by the presence of the rubber (i.e. a "solvent" effect), or accompanied by a second side reaction involving epoxides and a moiety on the rubber.

Esterification of an epoxy resin with 1300X8 (18% acrylonitrile) yields a material (theoretical EEW = 900 g eq^{-1} 10 wt% 1300X8) displaying a decreased rate of disappearance of epoxides during isothermal aging at 175C when compared with the 1300X13 material. A similar linear time dependent increase of EEW was observed for the analogous 1300X8 rubber-modified epoxy resin (Theoretical EEW = 900 g eq^{-1}, 10 wt% 1300X8). Table X summarizes the kinetic behavior and the composition of these resins.

Table X

Rate of Loss of Epoxides vs Elastomer Composition

EEW* (g eq^{-1})	Elastomer	$\frac{d[EEW]}{dt}$ (g eq^{-1} h^{-1})
900	none	8.8
900	10 wt% 1300X8	11.4
900	10 wt% 1300X13	14.5

* Theoretical EEW

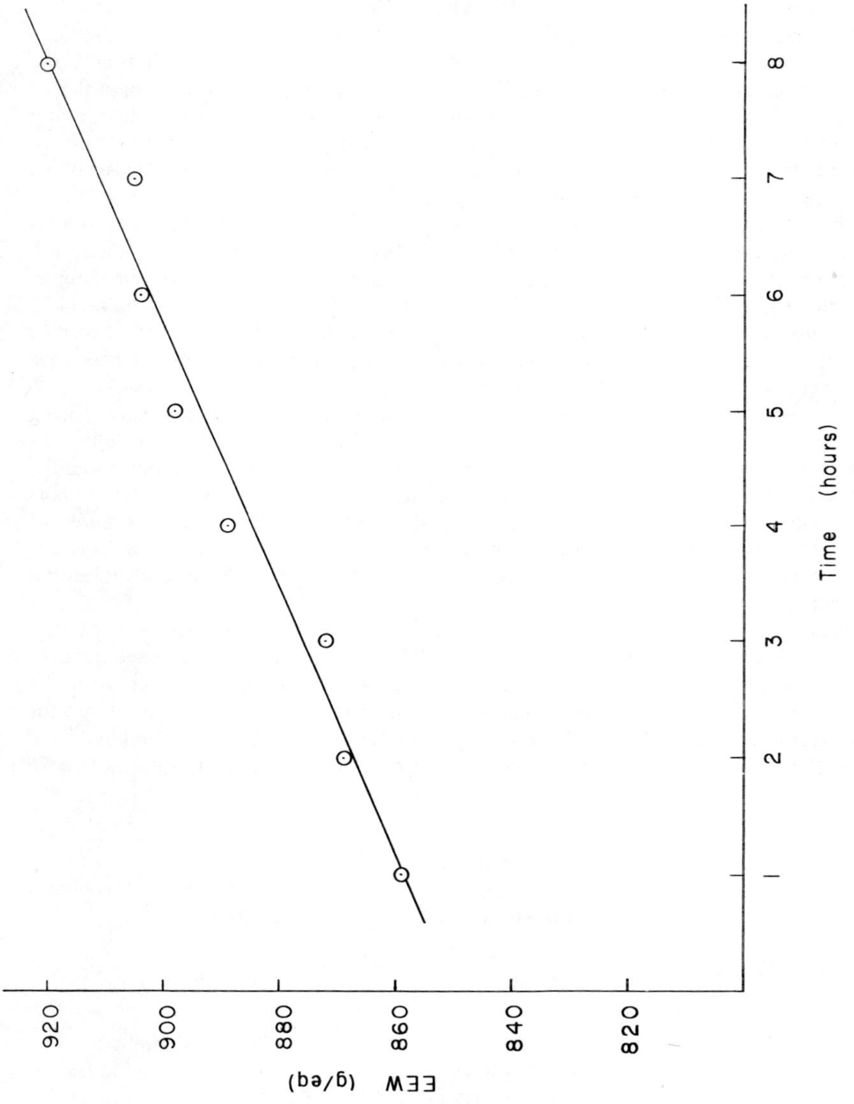

Figure 5. EEW vs Time for Nonmodified Epoxy.

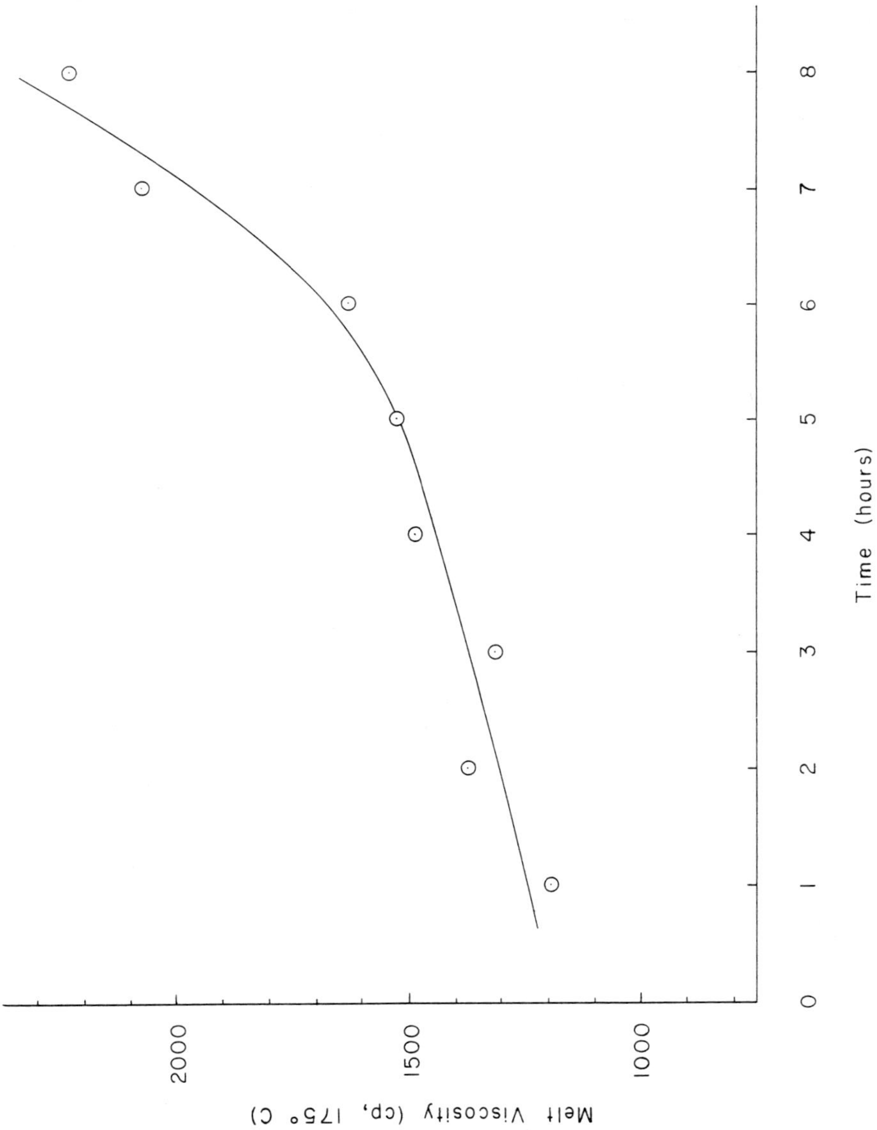

Figure 6. Melt Viscosity vs Time for Nonmodified Epoxy.

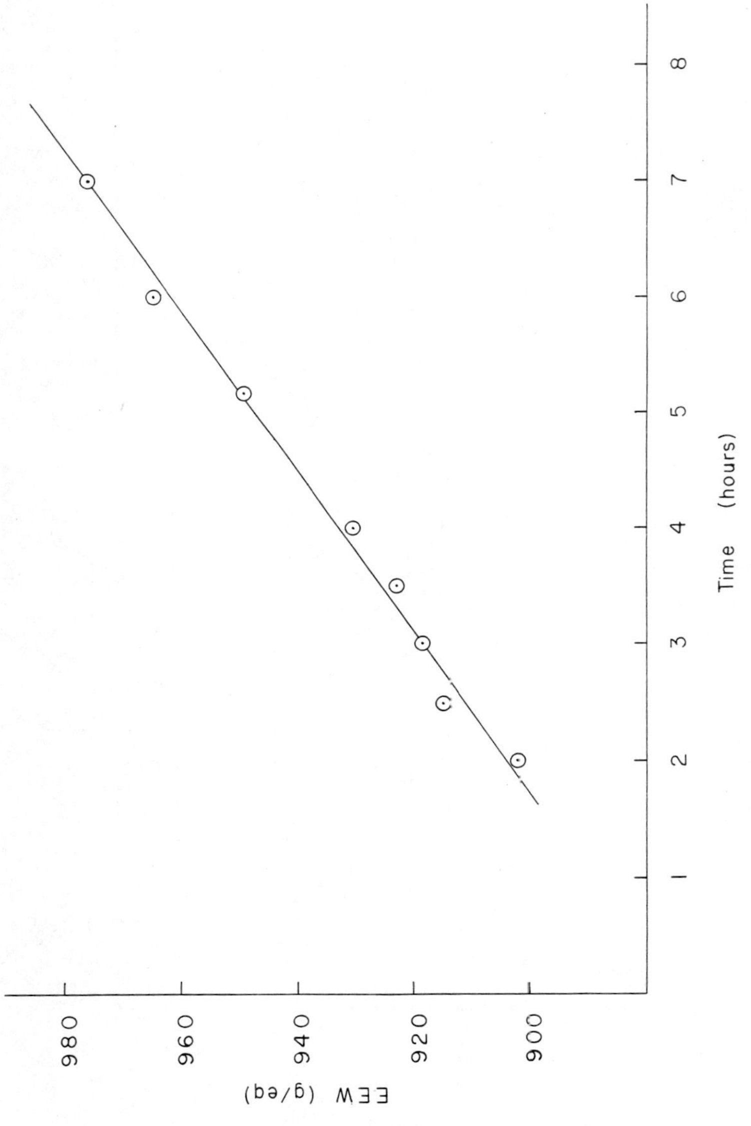

Figure 7. EEW vs Time for 10 wt% 1300X13 Modified Epoxy.

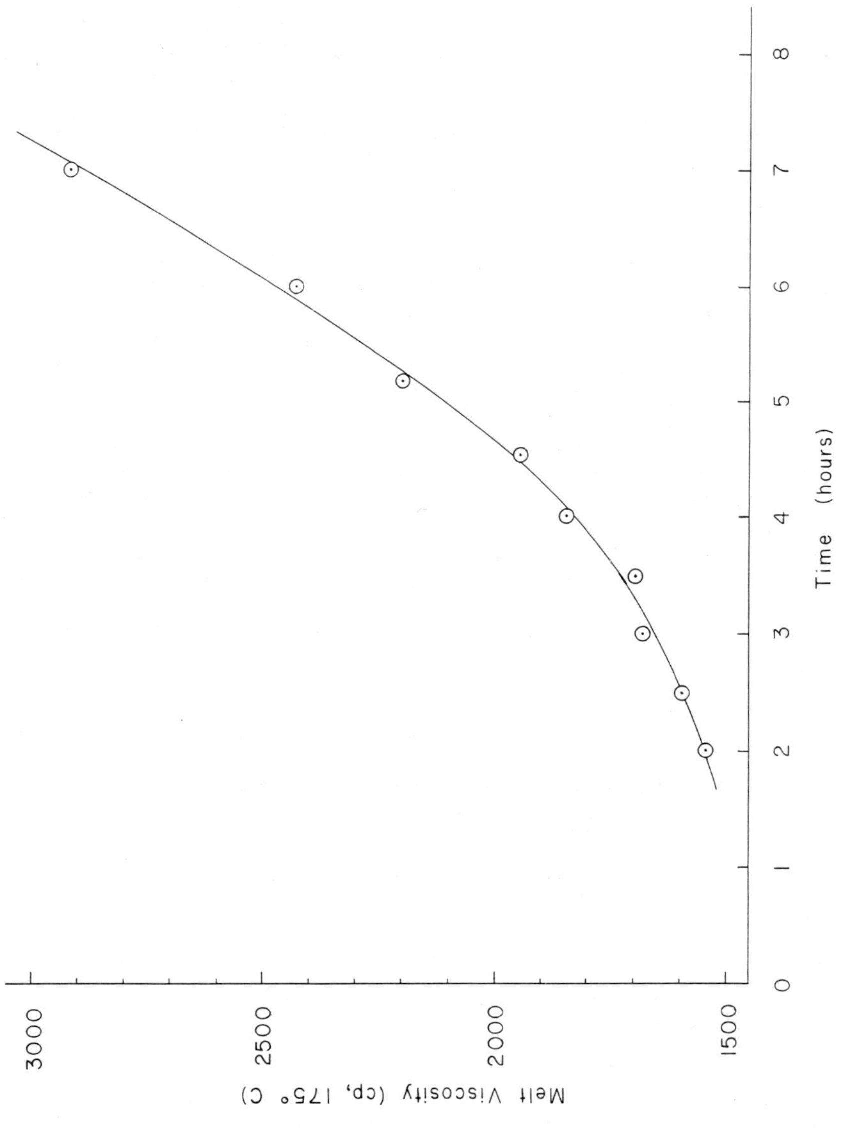

Figure 8. Melt Viscosity vs Time for 10 wt% 1300X13 Modified Epoxy.

The rate of loss of epoxides increases with the percent acrylonitrile in the CTBN elastomer. Figure 9 plots the rate of loss of epoxides versus percent acrylonitrile. These data, however, do not discriminate the two possible mechanisms for the loss of epoxides: 1. Enhancement of the rate of loss of epoxides via a "solvent" effect, and 2. A second side reaction involving reaction of epoxides and the rubber itself.

All of the above solid rubber-modified epoxy resins visually displayed clearly biphasic morphological properties (i.e., discreet rubber domains in a continuous epoxy matrix). If there is a reaction between a rubber moiety and an epoxide, it would best be studied in a homogeneous reaction mixture. Lower molecular weight epoxy resins are more compatible with CTBN elastomers and will form homogeneous solutions at elevated temperatures. Reaction of an epoxide with a reactive moiety contained in the elastomer, R, will most likely obey the following rate law:

$$\frac{-d[epoxide]}{dt} = k[epoxide][R] \qquad (5)$$

Thus, the disappearance of epoxides should be a function of both the epoxide concentration and the rubber moiety concentration. Since the various CTBN elastomers available span a range of stoichiometries of butadiene and acrylonitrile, esterification of these rubbers with liquid epoxy resin (EPON 828) provides a homologous series of liquid rubber-modified epoxy resins which will: 1. verify if the rate law in Equation (5) is satisfied, and 2. if Equation (5) is satisfied, identify which moiety on the rubber reacts with epoxides during isothermal aging at 175C. To enhance the observable rate of reaction, the concentrations of the reactive moieties were increased over those studied in the solid rubber-modified epoxy resins. The esterification of 50 wt% elastomer in EPON 828 was catalyzed at 130C by triphenylphosphine and the resulting product heated to 175C for isothermal aging. All esterifications went to greater than 99% completion as monitored by titration of the carboxylic acid groups on the elastomer. Table XI summarizes the compositions and isothermal aging behaviors for this homologous series of resins. Figures 10 - 13 show the EEW versus time at 175C for the 2000X162-, 1300X15-, 1300X8-, and 1300X13-modified liquid epoxy resins, respectively.

Table XI

Isothermal Aging vs Elastomer Composition
50 wt% Elastomer in EPON® 828

Elastomer	% Acrylonitrile	$\frac{d[EEW]}{dt}$ (g eq^{-1} h^{-1})
2000X162	0	2.0
1300X15	10	2.7
1300X8	18	2.8
1300X13	26	2.4

Although there is some scatter in the plots of EEW versus isothermal aging time, these data are convincing that, within experimental error, there is no significant difference in the rates of loss of epoxides between members of this homologous series of liquid rubber-modified epoxy resins. Further, that the rate of loss of epoxides is so low, despite the homogeneous reaction mixture and the high concentrations of epoxide

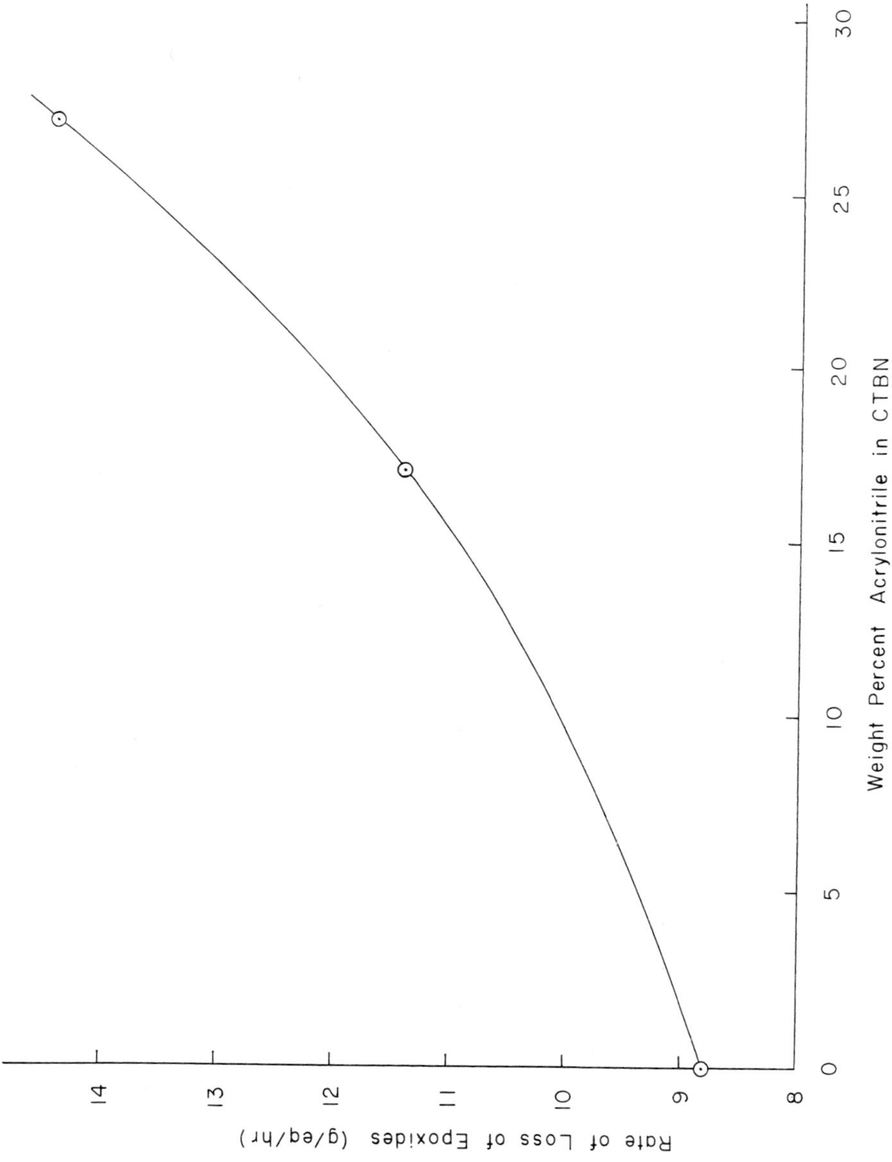

Figure 9. Rate of Loss of Epoxides vs Acrylonitrile Content.

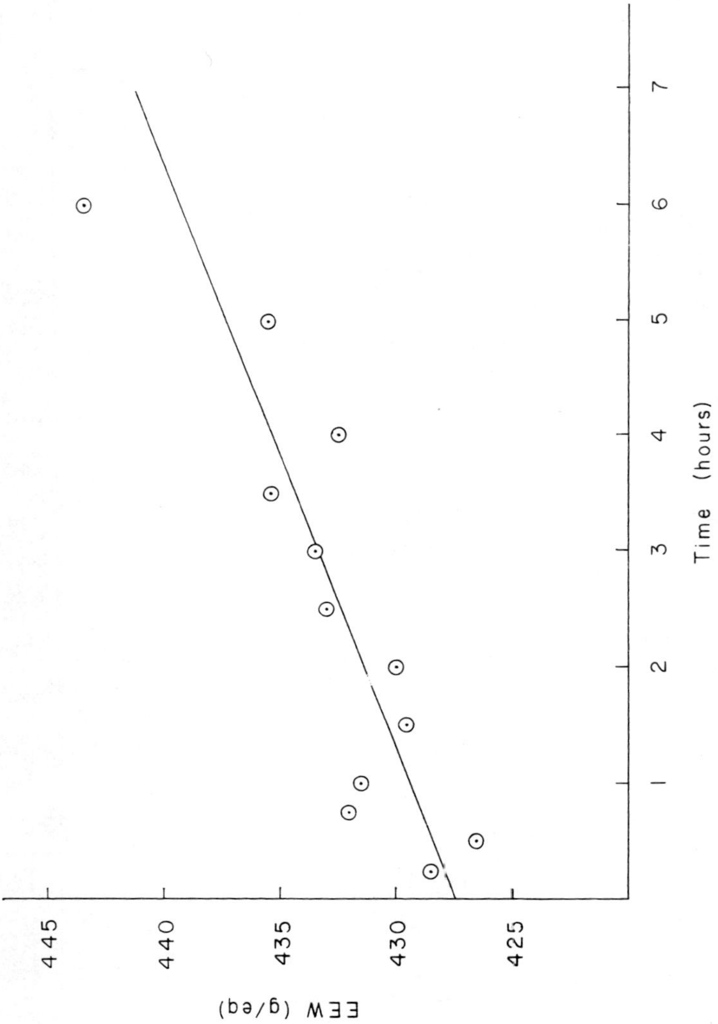

Figure 10. EEW vs Time for 50 wt% 2000X162 Modified Epoxy.

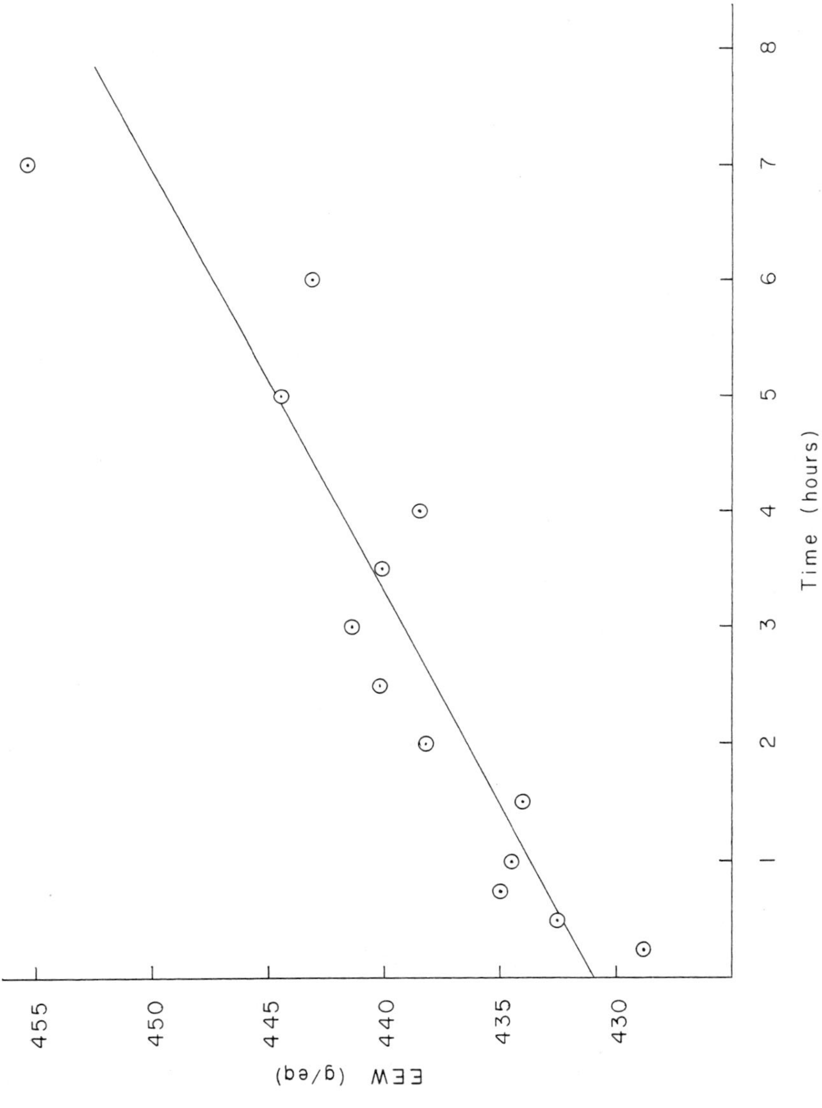

Figure 11. EEW vs Time for 50 wt% 1300X15 Modified Epoxy.

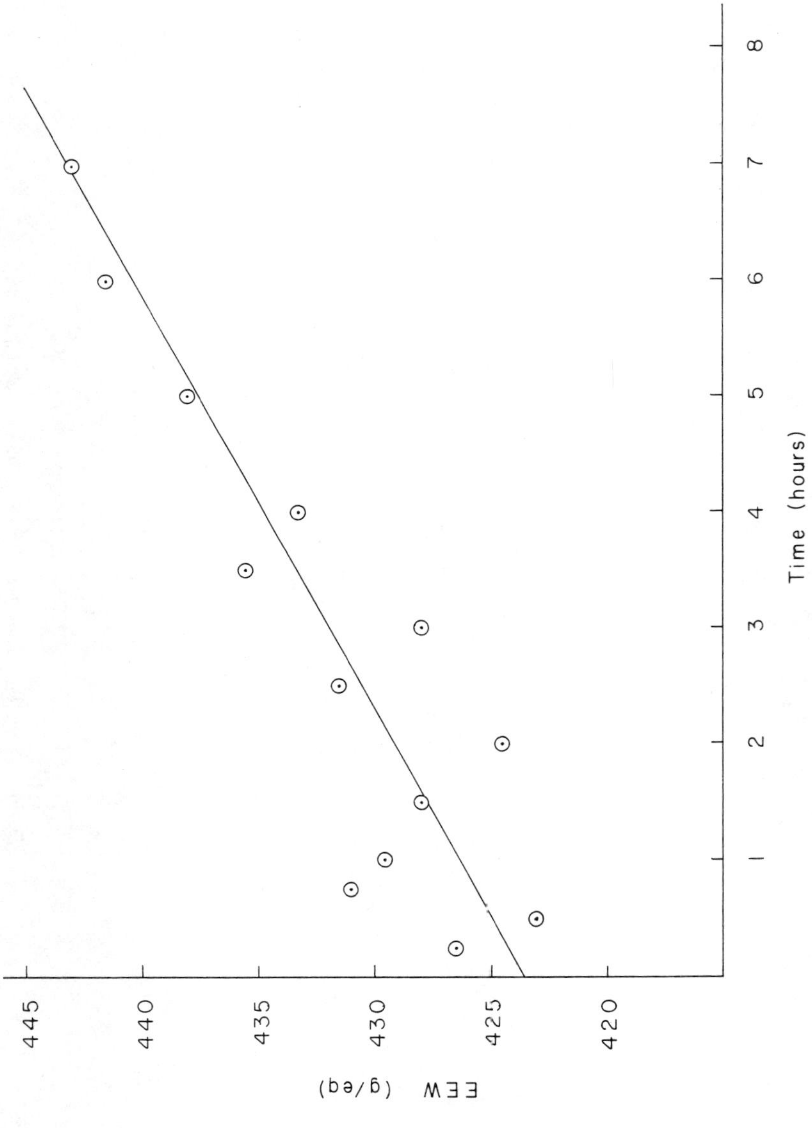

Figure 12. EEW vs Time for 50 wt% 1300X8 Modified Epoxy.

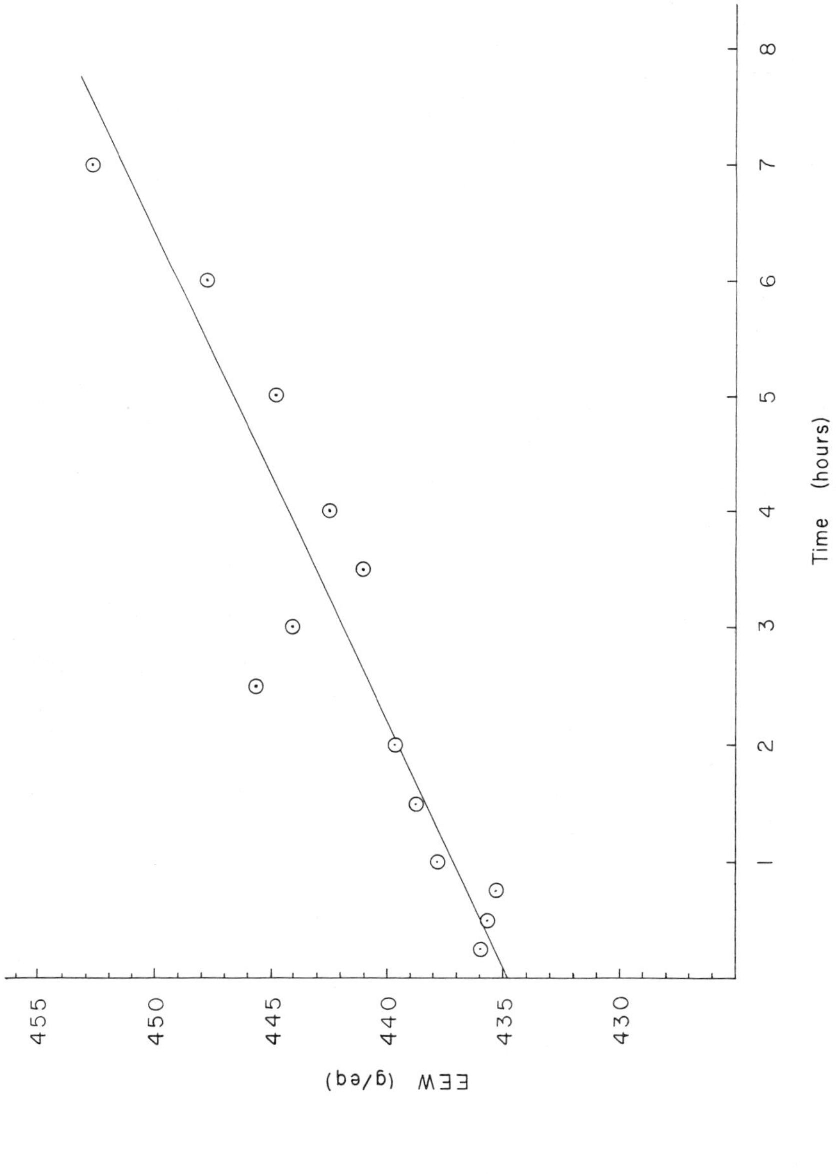

Figure 13. EEW vs Time for 50 wt% 1300X13 Modified Epoxy.

and elastomer, implies there is no specific reaction involving addition of epoxides to the backbone or pendant moieties on the elastomer. Thus, since Equation (5) is not satisfied, these findings support the premise that the elastomer enhances the rate of loss of epoxides via a "solvent" or reaction medium polarity-mediated effect.

The reaction which consumes epoxides yielding a higher viscosity resin presumably proceeds via reaction of the epoxide with a moiety on the epoxy resin backbone. This is consistent with the original observation that a non-modified epoxy resin undergoes this side reaction. Thus, this reaction ought to obey the rate law described in Equation (5) where R may be another epoxide, or the secondary alcohol.

The mechanism for the homopolymerization of epoxides may proceed via a bimolecular reaction yielding a symmetrical 2,5-disubstituted dioxane as shown in Reaction Scheme 6. This reaction should be described by the following kinetic expression for the disappearance of epoxides:

$$-\frac{d[epoxide]}{dt} = k[epoxide]^2 \qquad (6)$$

whose solution is

$$[epoxide] = kt + c \qquad (7)$$

A convenient form of the reciprocal of the epoxide concentration is the EEW (units are g eq^{-1}). Thus, merely plotting (EEW_t - EEW_o) versus time will yield a straight line whose slope is the bimolecular rate constant if this mechanism is operative. A second and more likely mechanism is shown in Reaction Scheme 7. The kinetics for disappearance of epoxides in this case will be described by a simple bimolecular expression:

$$-\frac{d[epoxide]}{dt} = k[epoxide][s-ROH] \qquad (8)$$

Inspection of Reaction Scheme 7 shows that the reaction of an epoxide with the secondary alcohol yields a branched molecule containing an ether and a secondary alcohol. This reaction consumes one secondary alcohol and yields one secondary alcohol. Thus, the concentration of secondary alcohol is a constant throughout the isothermal aging process. Therefore, the kinetic expression for the loss of epoxides simplifies to:

$$-\frac{d[epoxide]}{dt} = k^1[epoxide] \qquad (9)$$

where

$$k^1 = k[s-ROH] \qquad (10)$$

Equation (9) is a simple first order kinetic expression whose solution is:

$$\ln\left(\frac{[epoxide]_o}{[epoxide]_t}\right) = k^1 t \qquad (11)$$

For the concentration of epoxides, the inverse of EEW is used (e.g., for an EEW of 100 g eq^{-1}, the concentration is 0.001 eq g^{-1}). Thus, equation (8) becomes:

$$\ln\left(\frac{EEW_t}{EEW_o}\right) = k^1 t \qquad (12)$$

Similarly, the concentration of secondary alcohol is given by:

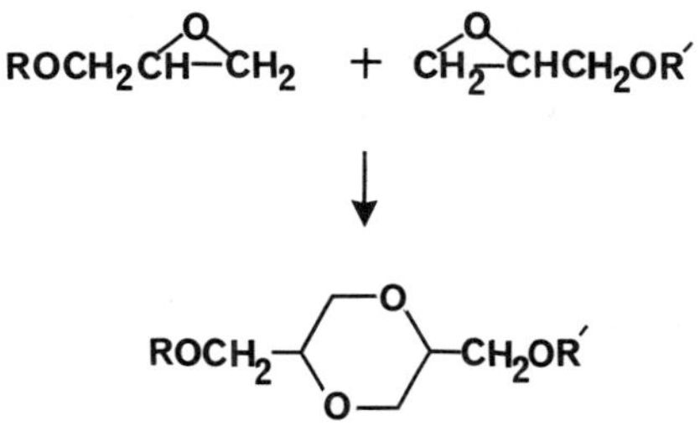

Reaction Scheme 6

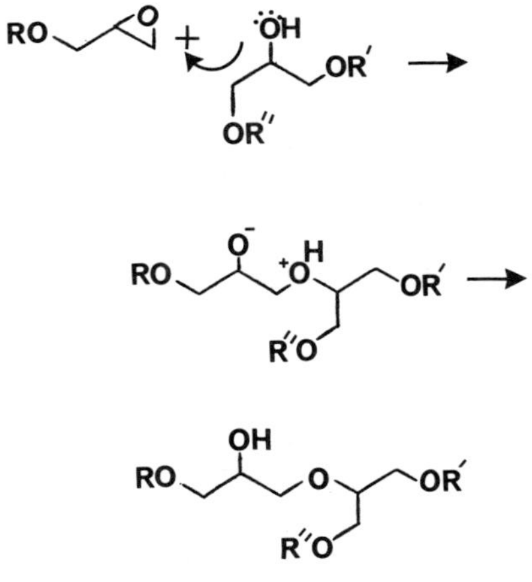

Reaction Scheme 7

$$[s{-}ROH] = [SAEW]^{-1} \quad (13)$$

where, SAEW is the secondary alcohol equivalent weight. The secondary alcohol equivalent weight may be conveniently calculated from the EEW. The pure epoxy monomer, the diglycidyl ether of bisphenol A, contains no secondary alcohols and has an EEW of 170 g eq^{-1}. Higher molecular weight epoxy resins are derived from advancement of the diglycidyl ether of bisphenol A with bisphenol A. This reaction consumes one equivalent of phenol for each equivalent of secondary alcohol in the final molecule (Reaction Scheme 1). The phenol equivalent weight of BPA is 114 g eq^{-1}. It can be derived that the secondary alcohol equivalent weight for a linear BPA-based epoxy resin is given by:

$$SAEW = \frac{100}{\left\{ \frac{\left[\left(\frac{100}{EEW} + \frac{100}{114}\right) x\ 114\right]}{\left(170 + 114\right)} - \frac{100}{EEW} \right\}} \quad (14)$$

Thus, determination of the EEW by titration allows calculation of the SAEW.

The methods are at hand to distinguish which mechanism is responsible for the increase in EEW during isothermal aging. The epoxide homopolymerization should have a second order dependence on epoxide concentration and no dependence on secondary alcohol concentration, whereas the epoxide-alcohol reaction should display a first order dependence on both epoxide and secondary alcohol concentration. Therefore, a study of isothermal aging kinetics versus EEW of the epoxy resin will distinguish these mechanisms.

An homologous series of epoxy resins with various EEWs was synthesized by standard advancement techniques and subsequently esterified with 15 wt% 1300X13. Isothermal aging kinetics were followed at 175C. Aliquots of resin were withdrawn hourly, rapidly cooled to room temperature, and titrated for EEW. No precautions were taken to exclude air during the isothermal aging. Table XII summarizes the formulations, reactive moiety equivalent weights, and kinetic calculations for the epoxide-alcohol addition reaction and the epoxide homopolymerization reaction.

Figure 14 shows the pseudo-first order kinetic plot for the loss of epoxides aging at 175C of a rubber modified epoxy resin (EEW_o = 375 g eq^{-1} before reaction with 15 wt% 1300X13). Figure 15 shows the same data treated for second order loss of epoxides. These data are still not sufficiently precise to rule out one mechanism on the basis of the linearity of the kinetic plots. However, the range of rate constants shown in Table XII for the epoxide-alcohol reaction kinetic treatment is significantly lower than for the epoxide homopolymerization kinetic treatment. If a reaction mechanism described by a rate law is operative, the rate constant obtained will not be a function of reagent concentration. The second order kinetic treatment for the homopolymerization of epoxides clearly does not meet this requirement, and can, therefore, be strongly discounted.

An additional mechanism resulting in an EEW overshoot of epoxy resins during isothermal aging was also discovered. The presence of oxygen influences the rate of overshoot. A rubber-modified epoxy resin (EEW = 1060 g eq^{-1}, 10 wt% 1300X13) was synthesized by standard advancement - esterification techniques. The solid resin was divided into aliquots and reheated to 180C under various reaction conditions. The resin was sampled upon reaching 180C, any required components were added, and the resultant mixture held isothermally at 180C for one hour, and thermally quenched to

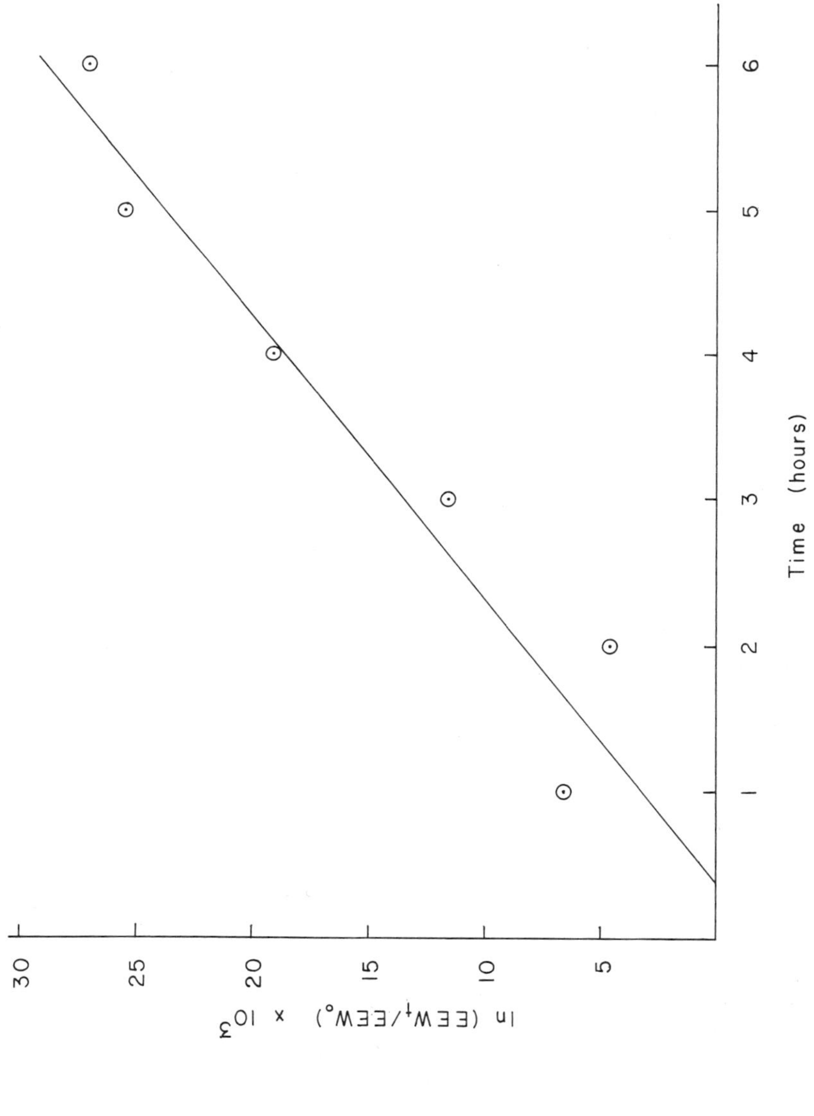

Figure 14. Pseudo-First Order Kinetics for Loss of Epoxides.

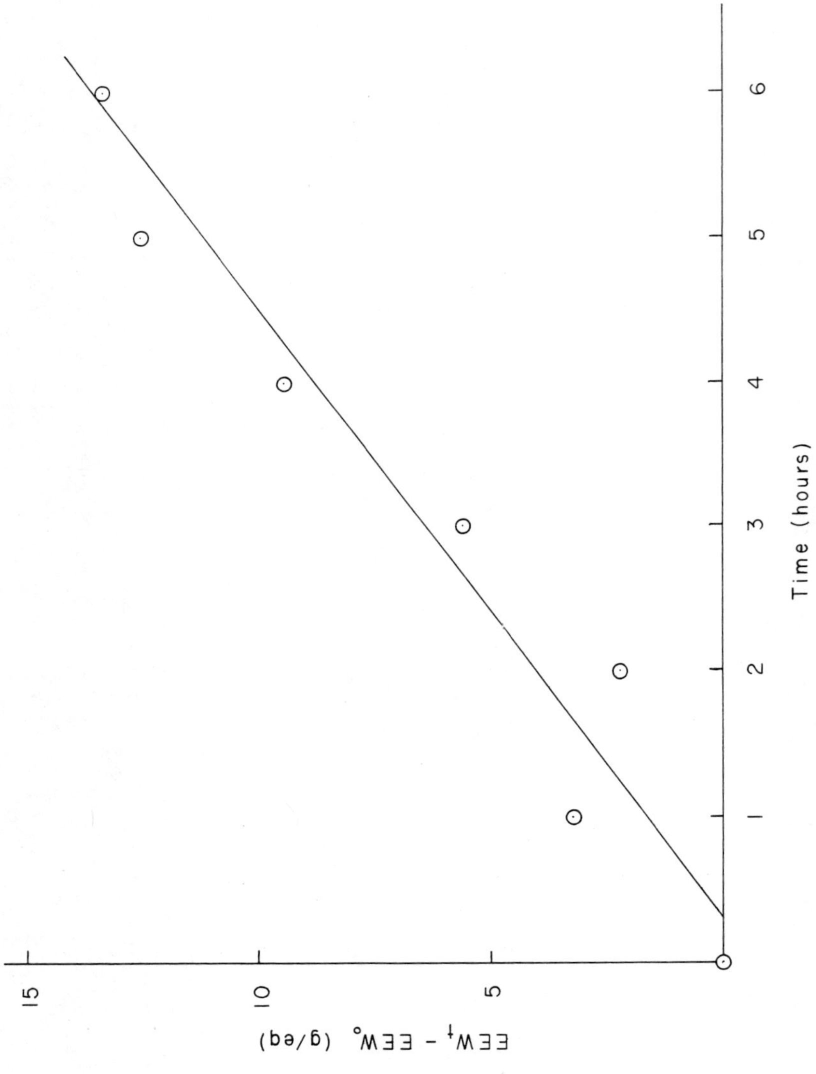

Figure 15. Second Order Kinetics for Loss of Epoxides.

Table XII

Isothermal Aging Kinetics for 15 wt% 1300X13 Modified Epoxy

EEW#	187	200	375
EEW_o	114	240	460
SAEW*	3652	2219	611
wt% EPON 828	85	83	69
wt% BPA	0	2	16
k^1 **	$4.8x10^{-7}$	$3.1x10^{-7}$	$19.9x10^{-7}$
k **	$17.4x10^{-4}$	$6.9x10^{-4}$	$8.5x10^{-4}$
k ***	$9.8x10^{-5}$	$7.7x10^{-5}$	$63.8x10^{-5}$

Prior to esterification.
* Calculation of SAEW was accomplished by adding a term to Equation (14) to account for the secondary alcohol produced in the esterification.
** Epoxide - alcohol reaction : pseudo first order kinetics (k^1: units of s^{-1}; k: units of g eq^{-1} s^{-1}).
*** Homopolymerization : second order kinetics (k units of g eq^{-1} s^{-1}).

room temperature. The reaction conditions investigated included the effect of additional triphenylphosphine, oxygen, and reaction kettle material (stainless steel versus glass). Table XIII summarizes the conditions studied and results obtained.

Table XIII

EEW Overshoot vs Reaction Conditions

Reaction	wt% Ph_3P	Atm.	Vessel	EEW_i*	EEW_f*
A	0	Air	Glass	1060	1115
B	0.05	Air	Glass	1060	1115
C	0.05	N_2	Glass	1060	1150
D	0.05	N_2	Stainless Steel	1090	1150

* Initial and final EEW.

Heating the resin to 180C had no effect on the EEW except in mixture D. This mixture was run in a larger beaker, thus necessitating a longer time to reach 180C. The higher initial EEW can be attributed to the longer heating period. Comparison of reactions A and B indicates no difference in the increase of EEW whether triphenylphosphine is added or not. This would lead one to conclude that the triphenylphosphine has no effect. The results of reactions C and D imply a probable occurrence with the triphenylphosphine. Note that in the absence of air (N_2 atmosphere) the magnitude of the EEW overshoot is larger. In the presence of air, the triphenylphosphine may be air oxidized at elevated temperatures, thus, no longer functioning to generate the betaine. Figure 16 shows a proposed life cycle for

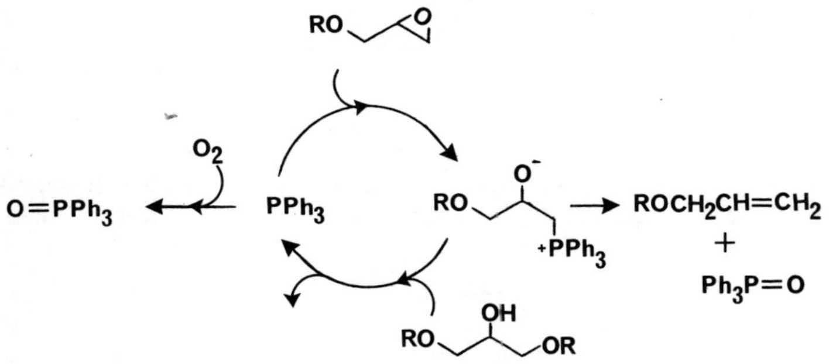

Figure 16. Proposed Life Cycle for Triphenylphosphine.

triphenylphosphine during isothermal aging. Note that the Wittig-type decomposition of the betaine results in triphenylphosphine oxide and a terminal olefin. The olefin cannot participate in crosslinking and, therefore, represents an undesirable biproduct associated with triphenylphosphine catalysis. Anaerobic conditions accentuate the EEW overshoot which implies that there may be a greater number of cycles through the loop; each cycle generates a branch point and the growth of higher molecular weight species. Alternatively, simple air oxidation of triphenylphosphine would eliminate the generation of terminal olefins, branch points, and the rapid loss of epoxides. It appears, therefore, that anaerobic conditions are not favored and, in fact, destabilize the epoxy resin product.

Conclusion A direct result of these studies was the successful scale-up of the synthesis of solid rubber-modified epoxy resins. It was shown that the presence of elastomers destabilizes epoxy resins such that the reaction between the terminal epoxide and pendant secondary hydroxyl groups is facilitated. Since the final product contains all of these moieties (i.e., elastomer, secondary alcohols, and epoxides) it is impossible to eliminate the undesirable side reaction. However, it is possible and advantageous to design the synthesis such that the magnitude of the hydroxyl-epoxide reaction is minimized.

Summary

The successful scale-up of advancement and modification of rubber-modified epoxy resins is discussed. Mechanisms are proposed for both advancement and esterification reactions as catalyzed by triphenylphosphine which are consistent with experimental results. A plausible mechanism for the destruction of the catalyst is also presented. The morphology of these materials is determined to be core-shell structures, dependent upon composition and reaction and processing conditions. Model studies have been performed to determine the effects of thermal history on the kinetics of reaction. These efforts have resulted in the successful scale-up and use of rubber-modified epoxy resins as functional coatings in the electronics industry.

Literature Cited

1. McGarry, F.J. and Willner, A.M., "Toughening of an Epoxy Resin by an Elastomeric Second Phase," Research Report R-68-8, School of Engineering, Massachusetts Institute of Technology, Cambridge, Mass., 1968.
2. McGarry, F.J., *Proc. Roy. Soc. Lond. A,* **1970,** *319,* 59.
3. Sultan, J.N. and McGarry, F.J., *Polym. Eng. Sci.,* **1973,** *13,* 29.
4. Laible, R.C. and McGarry, F.J., *Polym.-Plast. Technol. Eng.,* **1976,** *7,* 27, and references cited therein.
5. Meeks, A.C., *Polymer,* **1974,** *15,* 675.
6. Riew, C.K., Rowe, E.H., and Siebert, A.R., *ACS Adv. Chem. Ser.,* **1976,** *154,* 326.
7. Gillham, J.K., Glandt, C.A., and McPherson, C.A., "Characterization of Thermosetting Epoxy Systems Using a Torsional Pendulum," in "Chemistry and Properties of Crosslinked Polymers," S.S. Labana ed., Academic Press, New York, New York, 1977.
8. Bucknall, C.B. and Yoshii, T., *Brit. Polym. J.,* **1978,** *10,* 53, and references cited therein.
9. Manzione, L.T., Gillham, J.K., and McPherson, C.A., *J. Appl. Polym. Sci.,* **1981,** *26,* 889, 907, and references cited therein.
10. Wang, T.T. and Zupko, H.M., *ibid,* **1981,** *26,* 2391.

11. Sohn, J.E., *Am. Chem. Soc. Prepr., Div. Org. Coat. Plast. Chem.,* **1981,** *44,* 38.
12. Shechter, L. and Wynstra, J., *Ind. Eng. Chem.,* **1956,** *48,* 86.
13. Alvey, F.B., *J. Appl. Polym. Sci.,* **1969,** *13,* 1473.
14. Emerson, J.A. and Sohn, J.E., *Bull. Am. Phys. Soc.,* **1981,** *26;* and Roe, R.J., *J. Phys. Chem.,* **1967,** *71,* 4190.
15. Somerville, G.R. and Parry, H.L., *J. Paint. Tech.,* **1970,** *42,* 42.

RECEIVED December 2, 1982

6

Polyfunctional Chelating Agents for Improved Durability of Epoxy Adhesion to Steel

ANTHONY J. DeNICOLA, JR.[1] and JAMES P. BELL

University of Connecticut, Institute of Materials Science, Storrs, CT 06268

The serious drop in strength of epoxy bonded carbon steel structural joints exposed to humid environments for extended periods of time is greatly reduced by surface treating the bond surfaces with polyfunctional mercaptoester coupling agents. Adhesion performance was evaluated by accelerated aging of torsion joints in water at 57°C. Surface treatment involves preactivation of the steel surface with aqueous ammonium citrate solution and treatment with the mercaptoester in organic solution.

The durable adhesion of epoxy resins to steel in the presence of moisture has been shown to be difficult to achieve (1). The strength of these systems is typically adequate for most structural applications. However, the deleterious effect moisture has on the bonding between the metal and adhesive must be taken into account if epoxy resins are to be used for bonding steel in moist or humid environments. This places a severe limitation on the number of structural bonding applications where epoxide adhesives could be used.

Several research groups have studied the effects of water on the durability of structural joints. Kerr, MacDonald, and Orman (2) in their study of epoxy adhesives which were exposed to saturated atmospheres of water and ethanol found that hydrogen bonding interactions between the adhesive and metal were disrupted by the accumulation of water at hydrophilic sites on the oxide surface. The water displaces the resin where

[1] Current address: Hercules Incorporated, Hercules Research Center, Wilmington, DE 19899

0097-6156/83/0221-0119$06.00/0

secondary valence forces exist, thereby greatly weakening the bond. In contrast, it was found that ethanol, while affecting the strength of the joint by absorption into the matrix, had less tendency to adsorb at the interface and displace the resin from the surface. The authors suggest the tendency to aggregate at the interface is related to the capacity of the penetrant to hydrogen bond, water having a much higher capacity than ethanol. Similar studies confirm this concept of adhesive failure by preferential adsorption and adhesive displacement at the adherend surface (3).

Thermodynamic evidence for preferential displacement of the adhesive from the interface by water is provided by the work of Gledhill and Kinloch (1). By approximating the value of the work of adhesion between an epoxy adhesive and iron oxide surface, they were able to show that in the absence of water this value was positive and stable. If, however, water is present at the interface the work of adhesion shifts to a negative, and hence unstable value. It follows that under moist conditions, the locus of failure will likely be at the interface, resulting in a lower strength. This is what one observes experimentally.

From the above discussion, it follows that it should be possible to improve the durability of bonded joints by the introduction of suitable coupling agents at the interface. If the coupling agent is capable of chemically interacting with the metal or its oxide, displacement of the adhesive at the interface will be prevented.

Our efforts have focused on the possibility of attaching organic compounds onto the surface of 1018 low carbon steel through a complexation reaction with iron. We have examined mercaptoester chelating agents as potential coupling agents. These are compounds which are capable of complexing with metal ions by forming ring organometallic chelate structures.

Experimental. The adhesive employed was a diglycidyl ether of Bisphenol A (Epon Resin 828*) epoxide resin. The curing agents used to cure the epoxide resin and cure conditions are given in Table 1.

* Reg. trademark of Shell Chemical Co.

Table I. Amine Curing Agents and Conditions of Cure

AMINE	PHR, AMINE	CURE CYCLE	ACCELERATOR
** Versamid 140 Polyamide	50	72 hrs., 25°C	(0.2phr)Resorcinol
*** Diethylenetriamine	10	66 hrs., 25°C	--------
****DMP-30	10	1 hr., 120°C 2 hr., 150°C	--------
*** Methylenedianiline	25.6	1 hr., 120°C 2 hr., 150°C	--------

A tubular butt joint geometry developed by Lin and Bell (4,5) was used to test the shear strength of the epoxy/steel adhesive system. This geometry offers several advantages over some of the more conventional testing systems. The applied stress is very nearly simple shear. This eliminates geometric dependence of the data due to complex applied stresses. A second advantage is that the contact area between the adhesive and substrate is relatively small. The joint is designed so as to allow water penetration from both sides of a raised annular ring which is 0.16 cm in width. The maximum radial diffusion distance is 0.08 cm. This is the distance water would have to diffuse in order to completely cover the interface. Epoxy resin control systems (600 grit polished only) lose about 95% of their initial strength within a week when soaked in distilled water at 57°C. This allows accelerated wet strength testing without having to use extreme temperatures.

All joint surfaces were carefully polished with 600 grit sand paper. Joints were treated prior to applying the adhesive by totally immersing them in a refluxing organic solution containing the coupling agent for 15 minutes. After soaking in pure solvent for 5 minutes at room temperature, the joints were then air dried. The resin mixture (44 mg) was applied to the raised annular ring of the joint approximately one hour after treatment. The resin weight was measured accurately ($\pm$0.1 mg) by weighing the half of the joint containing the raised ring prior to and after coating. Both the adhesive joint assembly and steel sleeve placed around the joint to insure proper alignment are composed of 1018 carbon steel. Shear testing of the adhesive joints was conducted on an Instron Universal Testing Machine, Model TM-S. Torque was applied to the joint with an Instralab torsional device. A 1000-lb reversible load-cell was used to measure the force at break. The

** Reg. trademark of Henkel Corporation
*** Aldrich Chemical Co.
**** Reg. trademark of Rohm and Haas Company

cross-head speed was 0.2 in/min for all tests. Applied torque, Mt, is related to the maximum shear stress, tmax, by the following:

$$tmax = \frac{16.Mt.D_o}{\pi(D_o{}^4 - D_i{}^4)}$$

where D_i and D_o are the inner and outer diameters of the raised annular ring, respectively.

Coupling Agents

Hexanetriol trithioglycolate (HTTHG). 1,2,6-Hexanetriol (18 gm) and mercaptoacetic acid (37.5 gm) were added to 50 ml of xylene and heated to 150°C for 4.5 hours. Water formed during the reaction was azeotroped with xylene. Vacuum was applied at 160°C and 10 torr pressure for 1 hour to distill off excess reagents. The product was repeatedly washed with distilled water. Thirty-five ml of the product was isolated.

Ethylmercaptoacetate. Chloroacetylchloride (15 ml) was added to ethanol (100 ml). Potassium carbonate was added and the solution was stirred at room temperature for 1 hour. The potassium carbonate was filtered off. Excess ethanol was stripped on a rotary evaporator. The residue was added to 40 ml of toluene and the solution was washed with water. The toluene was stripped on an evaporator, yielding 15 ml of ethylchloroacetate.

Nash (20 gm) was dissolved in 200 ml of refluxing acetone. Ethylchloroacetate (10 ml) was added in two parts. After 3 hours of reaction toluene (400 ml) and water (200 ml) were added. The toluene layer was isolated and repeatedly washed with water. The solvent was stripped on an evaporator, yielding 8 ml of ethylmercaptoacetate.

Hexanetrioltrimercaptobutyrate (HTTMB). 4-chlorobutyrylchloride was added to 1,2,6-hexanetriol and the procedure for synthesizing ethylmercaptoacetate was followed.

Pentaerithritol tetrathioglycolate (PETG) and pentaerithritol tetramercaptoproprionate (PETMP) were provided by Evans Chemetics Company.

Results and Discussion. The curing agent used to crosslink an epoxy resin usually represents a significant fraction of the resin. In most situations it is the most important determining factor influencing the properties of the resin, since the vast majority of epoxy resins of commercial interest are based on bisphenol A. It is then not surprising to find that the curing agent can have a sizable effect on the durability of the system. Figures 1 and 2 illustrate the influence of curing agent on adhesive joint durability. All of the control systems studied rapidly lost their initial dry state properties when exposed to water. However, those systems which are either capable of shielding formed hydroxyl groups or which contain a low concentration of hydroxyl or other polar groups are much more resistant to the effect of water. This is the case for the Versamid 140 and DMP-30 cured epoxy systems. Figure 1 is a comparison of DETA and V-140 room temperature cured epoxies. The Versamid composition is much less susceptible to the effect of water, owing to the shielding ability of the fatty portions of the molecule (6).

Figure 2 illustrates that the same effect is present for high temperature cured epoxy systems. The DMP-30 composition is more durable than the methylenedianiline cured material because it involves polymerization of the epoxy group, forming a polyether network. Hence, the number of polar groups is reduced and the water durability of the adhesive joint is improved.

Figure 2 also points out another important point. Despite the very high shear strength achieved with methylenedianiline in the dry state, this system is actually the least durable of the four compositions examined. Dry strength adhesion is not a good measure of water durability.

The effect of hexanetriol trithioglycolate on the durability of the Versamid 140 cured system is presented in Figure 3. Treatment of the steel surface with this coupling agent results in a considerable increase in water durability. A previous study of B-diketone coupling agents performed in our laboratory (7) showed that a pre-activation of the surface was necessary for bond improvement. This pre-activation involves exposing the polished steel joints to a 3% aqueous ammonium citrate solution (pH 7.0) at 70°C prior to immersion in a methanol solution (2.5 x 10^{-3}M) of the coupling agent at reflux. A prior citrate pretreatment was used for all systems presented in this paper unless otherwise stated.

The effect of ammonium citrate treatment alone was examined and is presented in Figure 4. The use of citric acid provides short term improvement in durability for the Versamid cured system. There is a decay in strength to the level of the controls after prolonged soaking at 57°C.

Citric acid is known to be a strong chelating agent for metal ions (8). The iron complex of citric acid is water soluble. This drop in durability after prolonged soaking is thought to be due to the fact that citric acid forms a water soluble chelate with iron. Such a complex would be expected to be sensitive to the diffusion of water into the joint. It should be noted that all of the coupling agents investigated form water insoluble chelates.

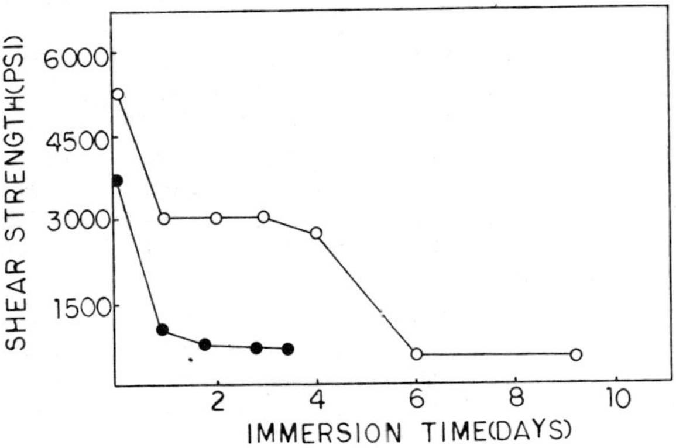

Figure 1.
The effect of water immersion (57°C) on the torsional shear strength of Versamid 140 (O) and DETA (●) room temperature cured adhesive joints.

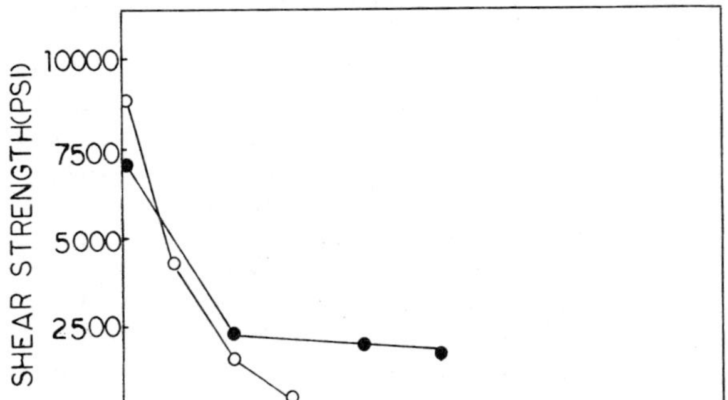

Figure 2.
The effect of water immersion (57°C) on the torsional shear strength of methylene dianiline (O) and DMP-30 (●) high temperature cured adhesive joints.

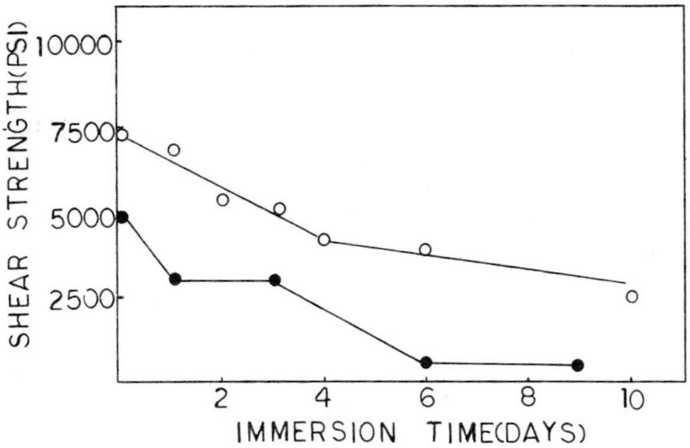

Figure 3.
Effect of HTTHG treatment on bond durability (57°C) of Versamid 140 cured joints: (O) HTTHG, (●) control.

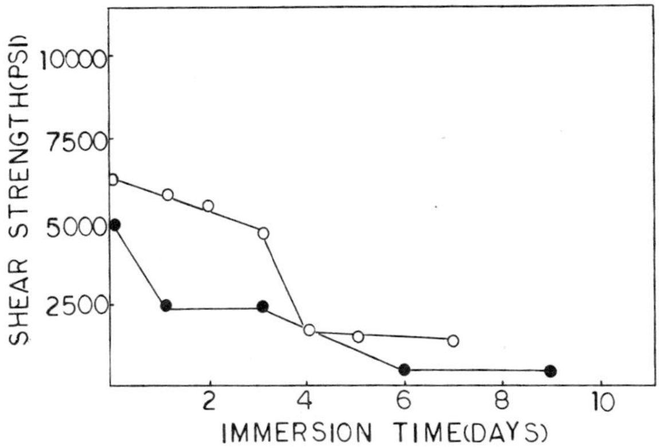

Figure 4.
Effect of ammonium citrate treatment on bond durability (57°C) of Versamid 140 cured joints: (O) ammonium citrate, (●) control.

The water absorption properties and shear strength of the Versamid 140 cured composition is presented in Figures 5 and 6, respectively. A 16 mil plate of resin was cured between glass plates. After curing it was cut into strips. Weight gain studies were performed by soaking the samples in distilled water at 57°C. The shear strength of the resin as a function of immersion time in water at 57°C was determined by a punch type shear device specified in ASTM D732-46.

Equilibrium water uptake is achieved in a very short time with thin samples. An equilibrium weight gain of 3.8% was achieved after approximately 48 hours at 57°C. A 33% loss of shear strength occurs after soaking in distilled water at 57°C. The shape of the curve (Figure 6) seems to indicate that this loss in strength is due to plasticization of the resin.

Figure 7 points out the tremendous improvement in water durability possible via the use of coupling agents. The methylenedianiline cured control falls to a strength level of 300 psi after 3 days in water at 57°C. The treated joints, on the other hand, maintain a strength of 3900 psi after 32 days in water.

Several factors are thought to be involved in the mechanism of durability improvement. The objective of this study was to examine compounds which are capable of interacting chemically with the steel surface. In addition, reaction with the epoxide resin is also desirable. The polyfunctional mercaptoester compounds meet both of these requirements in that the mercaptoester moiety provides a means of chelating iron cations and in addition is very reactive towards the epoxide group in the presence of an amine. The polyfunctional nature of this compound enables it to attach to the steel surface at more than one site and provides a group which can chemically attach to the resin. The dramatic improvement in bond durability supports such a hypothesis.

ESCA analysis of ammonium citrate and HTTHG treated steel is currently being undertaken in an attempt to better understand the chemical processes which are occuring at the surface. The following proposal is presented in an attempt to explain the bond improvement mechanism. As stated earlier, the ammonium citrate pretreatment was found to be a necessary component of the treatment. Citric acid is known to be an effective treatment for the removal of iron oxide from stainless steel (9), dissolving the oxide by forming a water soluble chelate with iron. Carbon steel (1018) was used in the present study. Citric acid dissolves the oxide film and replaces it with a thin film of citrate complex. Subsequent treatment with a mercaptoester coupling agent results in a replacement of the citrate by the mercaptoester, thereby forming a tightly bound insoluble complex layer on the surface. The citric acid, in

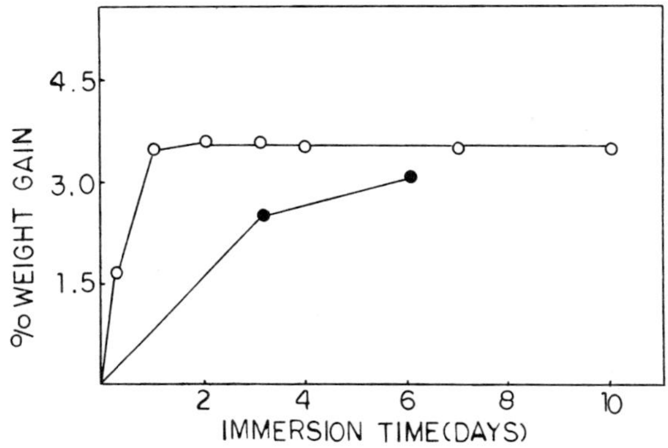

Figure 5.
% weight gain as a function of water immersion (57°C) of Versamid 140 cured epoxy: (O) 0.4 mm plate, (●) 1.6 mm plate.

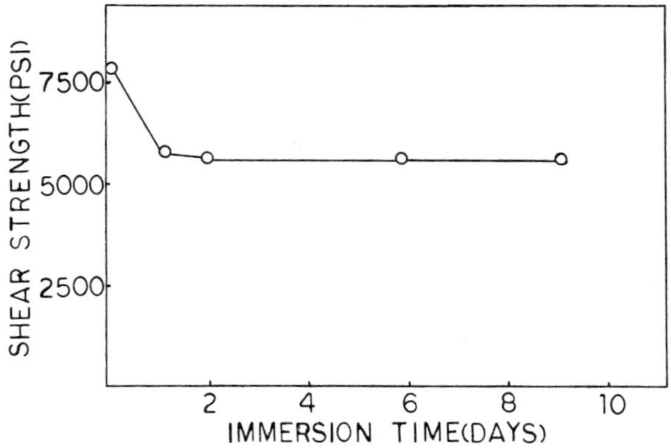

Figure 6.
Effect of water immersion on the shear strength of Versamid 140 room temperature cured epoxy (0.4 mm thick sample).

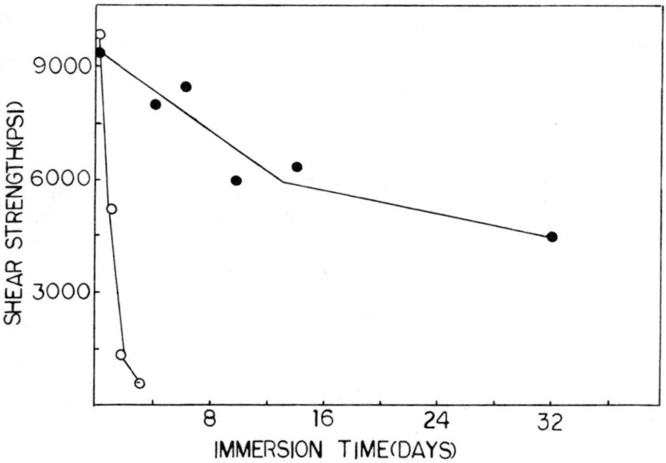

Figure 7.
Effect of HTTHG treatment on bond durability (57°C) of methylene dianiline cured adhesive joints: (O) HTTHG, (●) control.

effect, provides a source of iron cations. The coupling agent cannot effectively complex with iron in the oxide form. This is illustrated in Table II. When citrate is not used prior to treatment with the coupling agent, after 4 days immersion the strength of the MDA cured resin has dropped to 3800 psi. This compares with 7800 psi when citrate is used with the coupling agent.

Table II Effect of HTTHG Treatment on Joint Durability

Pretreatment: Test A - Treated with HTTHG in a refluxing solution (methanol) for 1/2 hour.
Test B - Treated with a 3% aqueous ammonium citrate (pH 7.0) solution. Subsequent treatment with HTTHG in methanol for 15 minutes.

	Shear Strength (psi)	
Joint No.	Test A	Test B
1	4300	8600
2	4600	7100
3	2600	7700
AVG:	3800	7800

1. Joints immersed in distilled water at 57°C for 96 hours.
2. Cured with methylenedianiline.

Additional evidence to support the concept that citric acid provides iron cations is presented in Table III. If ammonium citrate was acting only as a chemical oxide dissolver, then mechanical polishing under nitrogen would be expected to have the same effect. Mechanically polishing the steel joints and applying HTTHG under nitrogen resulted in a bond strength of 3500 psi after 240 hours immersion in water at 57°C. Polishing joints in air and then treating with ammonium citrate and HTTHG in nitrogen resulted in a bond strength of 7000 psi for the same immersion time. Activation of the surface by ammonium citrate is apparently involved.

Table III Comparison of Ammonium Citrate and Mechanical Polishing in Nitrogen

Pretreatment: Test A - Polished with 600 grit in a nitrogen atmosphere (0.3% O_2 max.)
Test B - Polished with 600 grit in air, then treated with a 3% ammonium citrate solution under nitrogen.

Joint No.	Shear Strength (psi) Test A	Test B
1	3900	8300
2	4200	6400
3	3000	6400
4	3100	
5	3400	
AVG:	3500	7000

1. All joints were treated with HTTHG in nitrogen.
2. O_2 concentration was determined with a mass spectrum.
3. Joints were immersed in distilled water at 57°C for 240 hours.
4. Cured with methylenedianiline.

Ammonium citrate solutions were found to lose their effectiveness when aged at room temperature for extended periods of time. Table IV illustrates the effect of aging of the ammonium citrate. Test specimens A were treated with ammonium citrate prepared one week prior to use. Test specimen B were treated with solution prepared just prior to use. All joints were subsequently treated with PETG in ethyl acetate, then immersed for 214 hours in water at 57°C. The bond strength of the fresh citrate treated joints is more than four times greater than the joints treated with aged citrate. The reason for this effect is not presently known, but perhaps it involves the formation of carbonate salts of the ammonium group via reaction with carbon dioxide in air. It is therefore important to use fresh solutions to obtain meaningful results.

Table IV Effect of Ammonium Citrate Aging on Joint Durability

Pretreatment: Test A - Treated with 3% aqueous ammonium citrate solution aged for one week in air. Test B - Treated with 3% aqueous ammonium citrate solution made just prior to use.

Joint No.	Shear Strength (psi) Test A	Test B
1	450	5140
2	1000	3950
3	900	5300
4	750	
5	750	
6	2500	
AVG:	1070	4800

1. All joints were treated with PETG in ethyl acetate.
2. Immersed for 214 hours in distilled water at 57°C.
3. Cured with methylene dianiline.

Pentaerithritol tetrathioglycolate (PETG) and pentaerithritol tetramercaptopropionate (PETMP) were examined as coupling agents to determine the effect of functionality and chelate ring size on bond durability. Both coupling agents provided a tremendous improvement in bond durability relative to the control. Figures 8 and 9 illustrate the effect of PETG and PETMP on bond durability for a methylenedianiline cured resin. The strength loss curves for these two materials resemble the HTTHG curve. Apparently, the strength loss process is common to all three coupling agents.

A comparison of the various surface treatments is presented in Table V. The resin used was methylenedianiline cured at high temperature. All of the joints were polished with 600 grit emery cloth and treated with ammonium citrate, with the exception of the control and sand blasted joints. Sandblasting the surface increases the surface area and provides a means of mechanical interlocking of the adhesive with the substrate surface. This does improve the water durability to some extent, but the effect is small when compared to the results achieved with the coupling agents. Ammonium citrate, in addition to activating the surface, has a beneficial effect on durability. It is conceivable that chemical reactions between the ammonium carboxylate groups of the citrate complex and the epoxy groups of the resin is occuring.

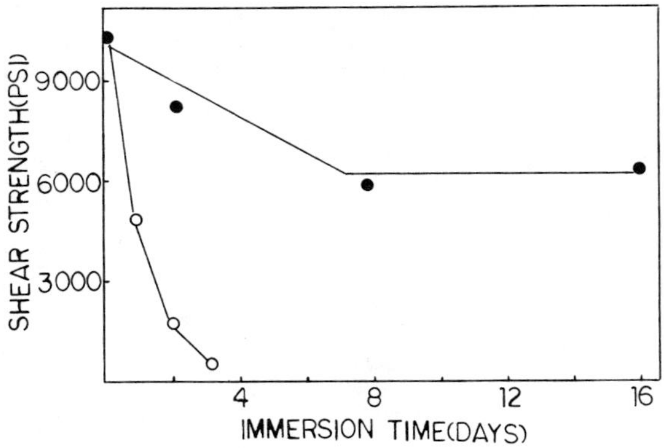

Figure 8.

Effect of PETG treatment on bond durability (57°C) of methylene dianiline cured adhesive joints: (O) PETG, (●) control.

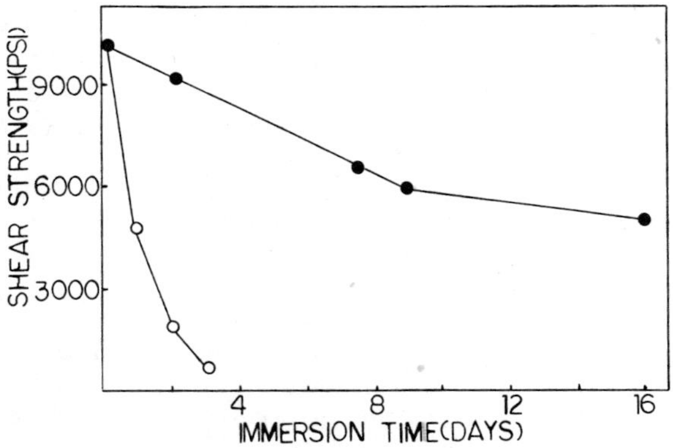

Figure 9.

Effect of PETMP treatment on bond durability (57°C) of methylene dianiline cured adhesive joints: (O) PETMP, (●) control.

Table V Comparison of the Various Surface Treatments

Pretreatment	Shear Strength (psi)
Control	0
Sandblasting	1560
Ammonium citrate	3900
Ethylmercaptoacetate	4580
HTTMB	7000
HTTHG	6525
PETG	6450
PETMP	6525

1. Joints were immersed in distilled water at 57°C for 240 hours.
2. Joints were polished with 600 grit emery cloth.
3. Joints were treated with the coupling agents after a preactivation with ammonium citrate.
4. Methylenedianiline cured.

Ethylmercaptoacetate, a monofunctional "coupling agent", provides a minor improvement relative to the citric acid, perhaps by providing a more hydrophobic surface. A decrease in functionality has a detrimental effect on bond durability relative to the polyfunctional coupling agents. Adequate coupling to both the substrate and resin are necessary for long term bond durability. Hexanetriol trimercaptobutyrate was examined to determine the effect of a seven memebered ring chelate on bond durability. The stability of a seven membered ring chelate would be expected to be lower than a five or six membered ring. This does not appear to be the case. Relative chelate dissociation does not correlate with bond strength. The same is true of functionality, provided a minimum functionality is present.

Conclusions. A tubular butt adhesive joint was found to be of particular advantage for studying the effects of water on the bond durability of epoxy adhesion to 1018 low carbon steel. This geometry lends itself to accelerated immersion testing without having to result to extreme temperatures.

Polyfunctional mercaptoester chelators have been synthesized and evaluated as potential coupling agents for the adhesion of epoxy resins to steel. The adhesive systems were evaluated by subjecting torsional joints to shear failure after accelerated immersion testing in distilled water.

The following results are of particular significance:

(1) The curing agent has an observable effect on bond durability- those curing agents capable of shielding polar

groups in the resin or forming crosslink structures not containing highly polar groups provide a higher degree of water durability.

(2) Pre-activation of the steel surface by ammonium citrate was necessary for maximum bond protection.

(4) Increasing the functionality of the coupling agent from three to four did not significantly alter the magnitude of the bond strength. However, reducing the functionality to one decreased the bond strength dramatically.

(5) Bond strength was not dependent upon chelate ring size.

Literature Cited

1. Gledhill, R. A.; Kinloch, A. J. J. Adhesion, 1974, 6, 315.
2. Kerr, C.; MacDonald, N. C.; Orman, J. J. Appl. Chem., 1967, 17, 62.
3. DeLollis, N. J.; Montaya, O. J. Appl. Poly. Sci., 1967, 11, 983.
4. Lin, C. J.; Bell, J. P. J. Appl. Poly. Sci., 1972, 16, 1721.
5. McCarvill, W. T.; Bell, J. P. J. Adhesion, 1974, 6(3), 185.
6. Witcoff, H. J. Oil and Colour Chem. Ass., 1964, 47(4), 273.
7. DeNicola, A. J.; Bell, J. P. Org. Coat. Plast. Chem. Preprints, Aug. 1981, 45, 120.
8. Blume, W. J. ASTM STP 538, 1973, 43.

RECEIVED December 2, 1982

7

High Performance Tris(hydroxyphenyl)methane-Based Epoxy Resins

K. L. HAWTHORNE and F. C. HENSON

Dow Chemical Company, Resins TS&D, Freeport, TX 77541

Epoxy resins based on the triglycidyl ether of tris(hydroxyphenyl)methane isomers and higher oligomers are multifunctional resins with improved thermal oxidative stability over other types of epoxy resins. They will help close the gap between phenol/formaldehyde novolac epoxies and higher temperature engineering plastics. When cured, these experimental tris(hydroxyphenyl)methane-based epoxies form a three-dimensional, tightly crosslinked structure with a high aromatic content and a very high heat distortion temperature. Initially, two experimental epoxy resins have been identified using this chemistry. Experimental Semisolid Epoxy Resin XD-7342.00L is aimed at high performance composites and adhesives where requirements include toughness, hot/wet strength and long-term high temperature oxidative resistance. Experimental Solid Epoxy Resin XD-9053.00L fits into the semiconductor molding powder industry where requirements include purity, formulated stability, fast reactivity and electrical properties over a broad temperature range.

Epoxy resins are versatile materials used in applications ranging from sand-filled floorings and two-part adhesives to aerospace composites and semiconductor encapsulation packages. They have gained acceptance over the past two decades by performing in many areas where other materials fail or are not cost-competitive. Properties like chemical resistance, toughness, adhesive strength, electrical insulation, heat and oxidation resistance and dimensional stability combine with a long list of processing techniques which made epoxy resins a 300 $\overline{M}$ lb/yr industry in the U.S. in 1980. Bisphenol A-based

0097–6156/83/0221–0135$06.00/0

epoxies account for the majority of this volume and perform well at temperatures up to slightly over 100°C. Novolac-based epoxy resins add utility at significantly higher temperatures. However, there is a considerable gap between high performance epoxy novolacs and other higher temperature materials such as silicones and polyimides which also carry much higher price premiums. The latest candidate to fill this need is a series of experimental epoxy resins based on the triglycidyl ether of tris(hydroxyphenyl)methane (patent pending) isomers and higher oligomers (from The Dow Chemical Company U.S.A.).

The trifunctionality of this resin results in a three dimensional, tightly crosslinked structure with a very high heat distortion temperature and improved thermal oxidative stability over other types of epoxy resins. Initially, two experimental epoxy resins have been identified using this chemistry and are currently manufactured in semicommercial quantities. Experimental Semisolid Epoxy Resin XD-7342.00L is aimed at advanced composites and adhesives where performance requirements include toughness, hot wet strength and long-term high temperature oxidative resistance. Experimental Solid Epoxy Resin XD-9053.00L fits into the semiconductor molding powders industry where resin requirements include purity, formulated stability, fast reactivity and excellent electrical properties over a broad temperature range.

This paper presents data generated to characterize the physical properties of tris(hydroxyphenyl)methane-based epoxy resins. Resin and cured clear casting physical properties, electrical properties, moisture resistance, formulated stability, reactivity and retention of properties at elevated temperature are covered compared to other selected resins.

Experimental

Liquid Physical Properties. Table I lists typical resin physical properties for XD-7342.00L and XD-9053.00L and the lower epoxide equivalent weight of XD-7342.00L agrees with its relatively lower viscosity. This suggests that XD-9053.00L contains a considerably larger fraction of higher molecular weight oligomers and that XD-7342.00L is predominantly the lower molecular weight triglycidyl ether of tris(hydroxyphenyl)methane. The lower viscosity of XD-7342.00L makes it suitable for adhesives and advanced composites where proper adhesive flow and composite reinforcement wetting become very difficult at high viscosities. The solid XD-9053.00L handles nicely in formulating semiconductor molding compounds where a nonsintering, crushable solid resin handles best. Table I also shows the higher purity of XD-9053.00L. Purity is a must for semiconductor encapsulation where contaminants combine with moisture and corrode

TABLE I

TYPICAL LIQUID RESIN PROPERTIES
OF
EXPERIMENTAL EPOXY RESINS XD-7342.00L AND XD-9053.00L

PROPERTY	XD-7342.00L	XD-9053.00L
EPOXIDE EQUIVALENT WEIGHT	162	220
SPECIFIC GRAVITY, 25°C	1.22	1.22
VISCOSITY (CENTISTOKES)		
150°C	55	550
60°C	14,000	-
DURRAN'S SOFTENING POINT, °C	55	85
VOLATILES, WT. % MAX.	1.0	0.5
HYDROLYZABLE CHLORIDES, WT. % MAX.	0.1	0.05
IONIC CHLORIDE, PPM MAX	10	5
SODIUM, PPM MAX.	15	5

parts of the electronic device which leads to premature device failure. Extremely high purity is not so critical in composites and adhesives unless the contaminants begin to interfere with cure which typically happens at much higher levels than those reported in Table I.

Clear Casting Physical Properties. Unreinforced clear casting physical property data using nadic methyl anhydride as a curing agent is in Table II. Immediately evident is the increase in heat distortion temperature (HDT) using the tris(hydroxyphenyl)methane-based resins compared to both phenol and cresol-based novolac epoxies. XD-9053.00L and XD-7342.00L also maintain a higher percentage of their room temperature strength at elevated temperature. Electrical properties at ambient and elevated temperatures measured according to ASTM-D257 and ASTM-D150 are also reported in Table II. As one can see, tris(hydroxyphenyl)methane-based epoxy resins maintain electrical insulation properties over a wide range of temperatures. Table III lists clear casting data for XD-9053.00L compared to solid cresol epoxy novolac resin using a solid phenolic curing agent. A trend similar to the anhydride-cured systems in the preceding table is evident. XD-9053.00L has a higher HDT and a correspondingly greater percent retention of physical properties at elevated temperature. Electrical properties of both systems are comparably good over the temperature range tested; however, the greater strength retention of XD-9053.00L should allow its use at considerably higher service temperatures.

Thermal Resistance. Additional technical data was gathered using several Du Pont instruments including: Thermogravimetric Analyzer (TGA), Differential Scanning Calorimeter (DSC) and Dynamic Mechanical Analyzer (DMA). The samples studied were tetraglycidyl methylenedianiline (TGMDA), XD-7342.00L and blends of XD-7342.00L and D.E.R. (Trademark of The Dow Chemical Company) 332. All the systems were cured with diaminodiphenyl sulfone (DADS) accelerated with boron trifluoride monoethylamine (BF_3·MEA) complex. The DADS was blended in at 130°C, and the BF_3·MEA was added after the resin had cooled to 100-110°C. The oven temperature was then reduced to 55°C for 4 hours followed by 125°C for 2 hours and finally 175°C for 2 hours. The test samples in the DSC were heated up to 295°C at 20°C/minute for the first scan. The sharp increase in heat flow (Figure 1) from the TGMDA system beginning at about 240°C indicates this system was not as completely cured as the XD-7342.00L system, even though they both received the same cure schedule. This may indicate a problem with dimensional stability using TGMDA relative to XD-7342.00L which exhibits a much smaller exotherm.

In the second scan (Figure 2) the same samples were

TABLE II

TYPICAL CLEAR CASTING PHYSICAL PROPERTIES
OF
RESINS CURED WITH NADIC METHYL ANHYDRIDE (NMA)[1]

PROPERTY	XD-7342.00L	XD-9053.00L	D.E.N. 438	EPOXY CRESOL NOVOLAC
HEAT DISTORTION TEMP., °C	236	223	183	189
FLEX. STRENGTH, PSI				
ROOM TEMP.	12,800	19,000	15,600	18,300
100°C	12,200	13,000	14,400	15,200
150°C	9,100	12,400	8,000	10,600
180°C	8,100	10,300	5,500	6,000
220°C	6,900	-	-	-
260°C	2,900	-	780	-
FLEX. MODULUS, PSI				
ROOM TEMP.	4.9×10^5	5.1×10^5	4.8×10^5	5.6×10^5
150°C	-	3.3×10^5	-	2.9×10^5
180°C	-	3.0×10^5	-	1.6×10^5
DISSIPATION FACTOR				
ROOM TEMP.	0.007	0.006	0.006	0.006
150°C	0.007	0.003	-	0.004
180°C	0.009	0.003	-	0.007
DIELECTRIC CONSTANT				
ROOM TEMP.	3.8	3.7	3.4	3.4
150°C	3.9	3.8	-	3.5
180°C	3.9	3.8	-	3.5
VOLUME RESISTIVITY, OHM cm				
ROOM TEMP.	$>10^{15}$	$>10^{15}$	$>10^{15}$	$>10^{15}$
150°C	6.8×10^{11}	1.1×10^{12}	-	1.2×10^{12}
180°C	3.7×10^{11}	1.9×10^{11}	-	4.3×10^{11}

[1]85% OF STOICHIOMETRIC AMOUNT OF NMA ACCELERATED WITH 1.5 PHR (PARTS PER HUNDRED PARTS OF RESIN) BENZYLDIMETHYLAMINE

TABLE III

TYPICAL CLEAR CASTING DATA OF PHENOLIC-CURED XD-9053.00L COMPARED WITH EPOXY CRESOL NOVOLAC RESIN

PROPERTY	XD-9053.00L	EPOXY CRESOL NOVOLAC
HEAT DISTORTION TEMPERATURE, °C	213	154
FLEXURAL STRENGTH, PSI		
ROOM TEMPERATURE	16,600	21,100
100°C	14,400	15,400
150°C	12,500	5,800
180°C	7,600	1,200
FLEXURAL MODULUS, PSI		
ROOM TEMPERATURE	5.0×10^5	5.4×10^5
100°C	3.7×10^5	4.4×10^5
150°C	3.0×10^5	1.5×10^5
180°C	1.9×10^5	0.1×10^5
DISSIPATION FACTOR		
ROOM TEMPERATURE	0.004	0.005
150°C	0.004	0.005
180°C	0.012	0.015
DIELECTRIC CONSTANT		
ROOM TEMPERATURE	4.2	3.9
150°C	4.2	3.9
180°C	4.2	4.1
VOLUME RESISTIVITY, OHM cm		
ROOM TEMPERATURE	$>10^{15}$	$>10^{15}$
150°C	2.3×10^{11}	4.7×10^{11}
180°C	3.1×10^{10}	4.0×10^{10}

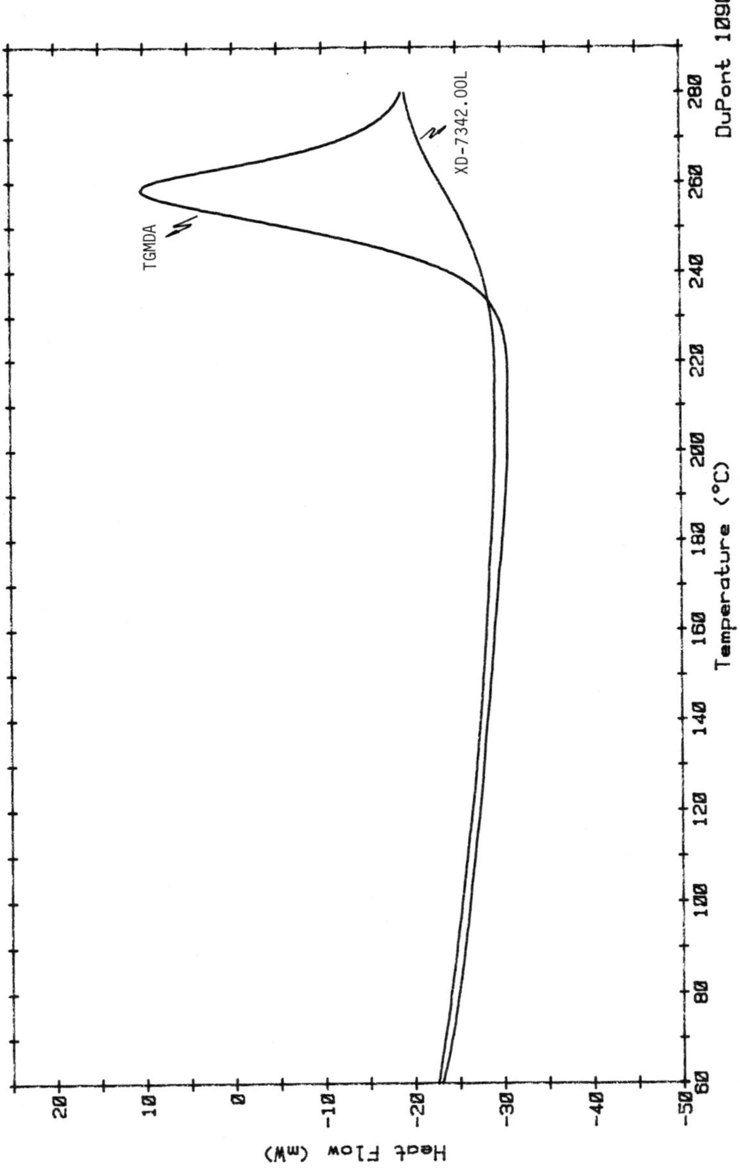

Figure 1. Differential scanning calorimeter; first scan is DADS/BF_3·MEA CURED.

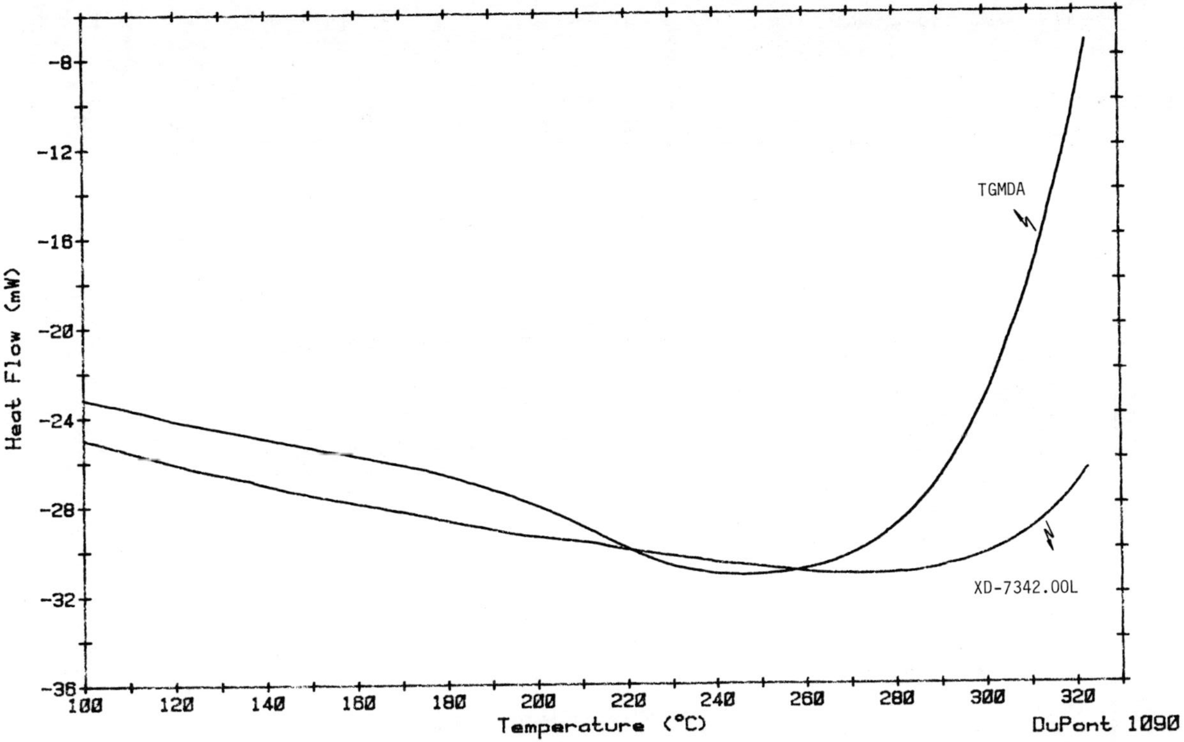

Figure 2. Differential scanning calorimeter; second scan is DADS/BF_3·MEA CURED.

heated at the same rate but to a higher temperature. The TGMDA scan indicates heat being given off (probably pyrolysis and thermal decomposition) at around 270°C, while the XD-7342.00L-based system reaches 300°C before giving off heat at a similar rate as evidenced by the slope of the DSC scans.

Thermogravimetric analysis (TGA) measures thermal stability by monitoring sample weight loss as a function of time and temperature. Figure 3 illustrates the superior thermal oxidative stability in air of XD-7342.00L compared to TGMDA. Using 5% weight loss as an indication of significant thermal damage, the TGMDA sample hits this point at 325°C, while XD-7342.00L went to 385°C before showing the same weight loss. It was noted that at around 360°C the TGMDA sample began to bubble and expand irregularly into a friable crust with a high void content. While the XD-7342.00L sample eventually lost considerable weight, it remained intact and retained its original shape.

Figure 4 is a Dynamic Mechanical Analyzer scan of XD-7342.00L cured with DADS/BF_3·MEA. The DMA measures the resonance frequency of a resin sample. This frequency changes with temperature and drops sharply at the sample's glass transition temperature. At the same time, the damping of the vibrational frequency through the sample peaks as the resonance frequency drops off. Using the peak of the damping curve as an indication of glass transition temperature (Tg), Figure 4 shows a value of 310°C for XD-7342.00L. Figure 5 is a DMA scan of XD-7342.00L diluted with 20% by weight D.E.R. 332 (bisphenol A-based epoxy) and its damping peak is at 307°C. Both these values are much superior to TGMDA which has a damping peak at 269°C shown in Figure 6. The cure system and postcure schedule for these samples were the same as previous DADS/BF_3·MEA-cured systems.

Reactivity and Stability. In the electronics industry both stability of a formulated molding compound and its reactivity are important. Good stability allows storage and transportation with minimal refrigeration and fast reactivity increases molding throughput. Table IV and Figure 7 illustrate several characteristics of XD-9053.00L compared to an epoxy cresol novolac with regard to these properties. From Table IV the data shows the XD-9053.00L-based system reacts 10% faster than the epoxy cresol novolac system measured by stroke cure at 160°C. Correspondingly, the XD-9053.00L system develops almost twice the Shore "D" hardness after 2½ minutes at 150°C compared to the cresol epoxy system. Even though the initial spiral flow of the XD-9053.00L-based compound is less than that of the cresol epoxy novolac, Figure 7 illustrates its superior formulated stability after 12 days at 40°C maintaining 67% of its original flow compared to 23% for the cresol epoxy novolac.

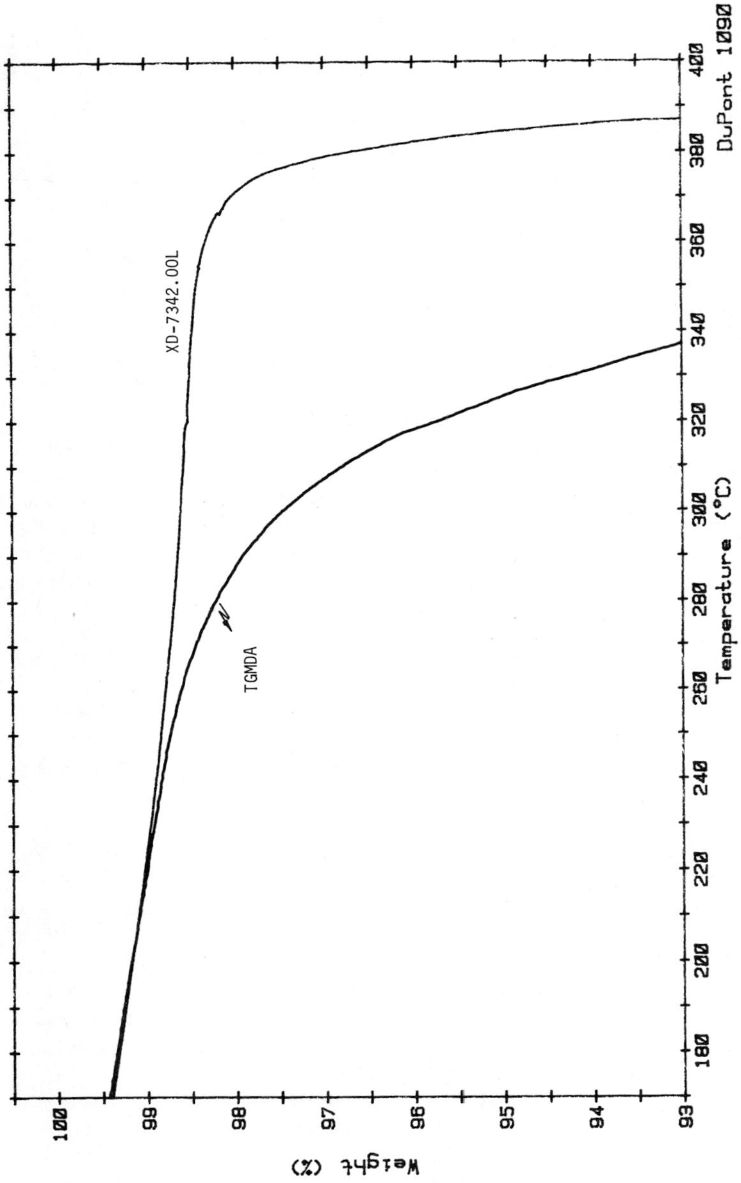

Figure 3. Thermogravimetric analysis. Conditions: 10 °C/min, DADS/BF_3·MEA CURED.

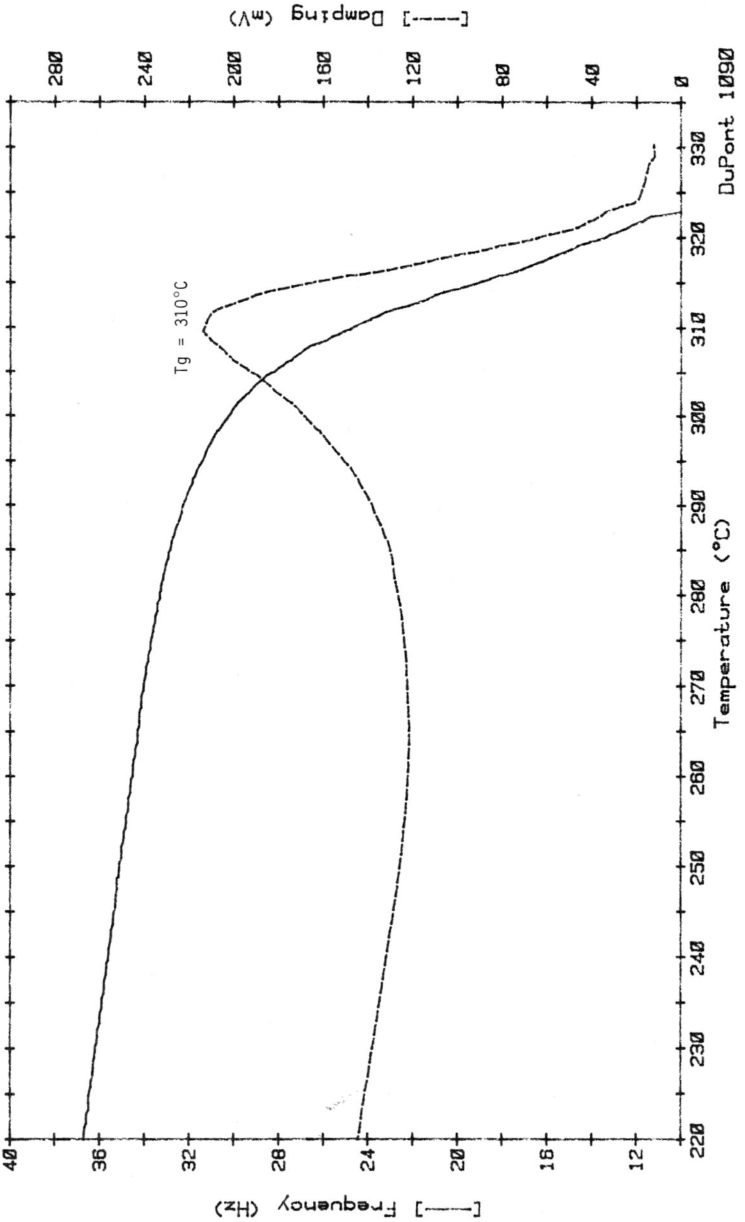

Figure 4. Dynamic mechanical analysis. Conditions: 2 °C/min, XD-7342.00L, DADS/BF_3·MEA CURED.

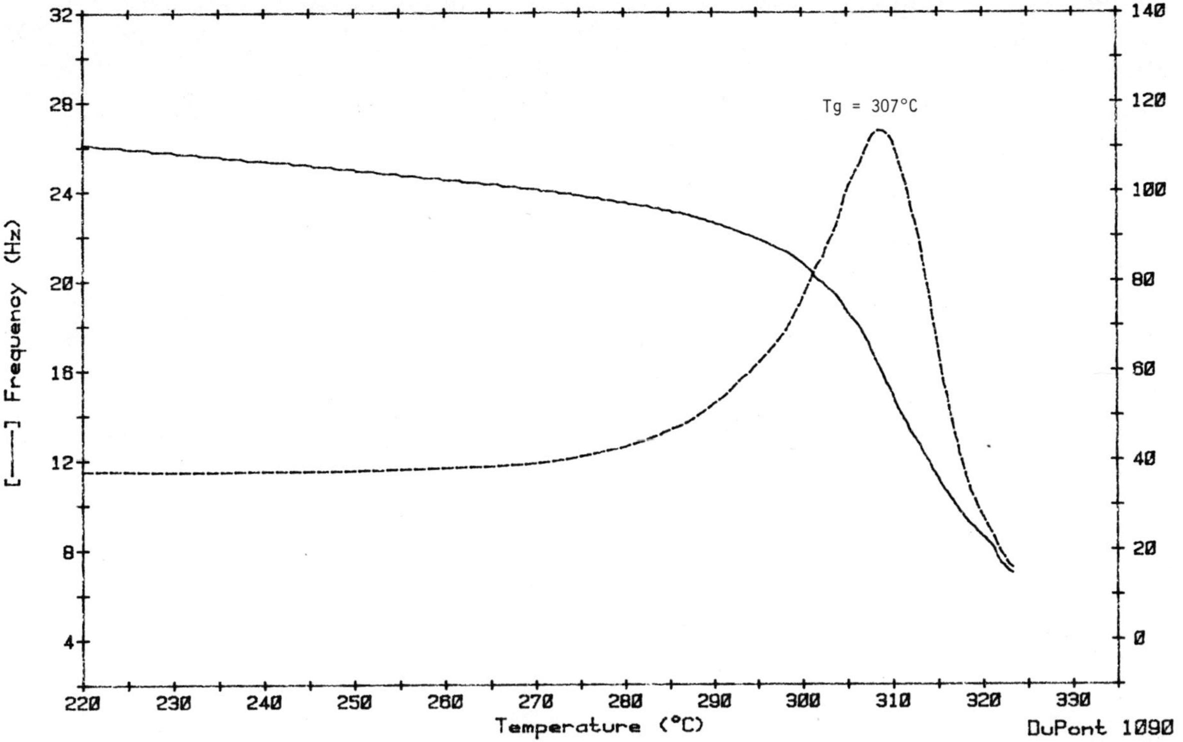

Figure 5. Dynamic mechanical analysis. Conditions: 5 °C/min, 80% XD-7342.00L, 20% D.E.R.* 332, DADS/BF_3·MEA CURED.

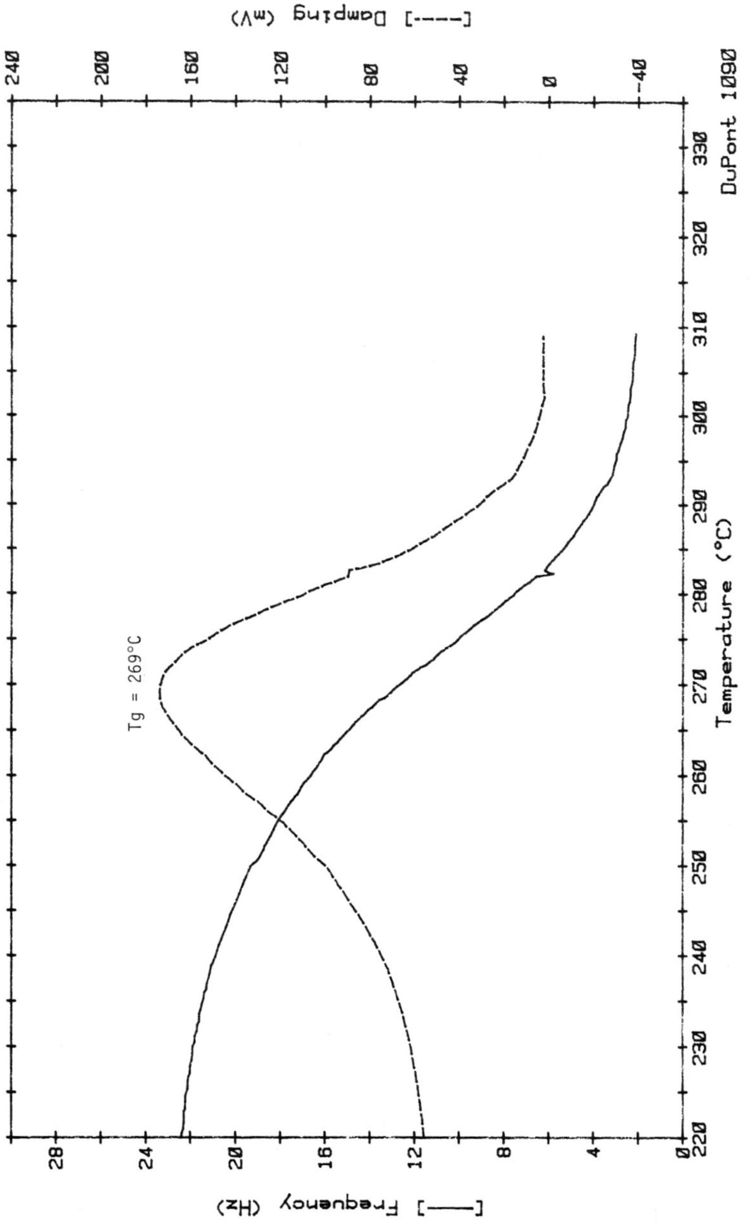

Figure 6. Dynamic mechanical analysis. Conditions: 2 °C/min, TGMDA, $DADS/BF_3{\cdot}MEA$ CURED.

TABLE IV

REACTIVITY OF TRANSFER MOLDING COMPOUND MADE USING XD-9053.00L VS. CRESOL EPOXY NOVOLAC*

	XD-9053.00L	EPOXY CRESOL NOVOLAC
ORIGINAL FLOW EMMI, INCHES	47	53
STROKE CURE, SECONDS @ 160°C	60	67
HOT HARDNESS, SHORE "D" 2.5 MINUTES @ 150°C	38	20

*PHENOLIC-TYPE CURING AGENT ACCELERATED WITH 0.45 PHR METHYL-IMIDAZOLE AND CONTAINING 67 WT. % SILICA FLOUR AND 5 PHR MONTAN WAX

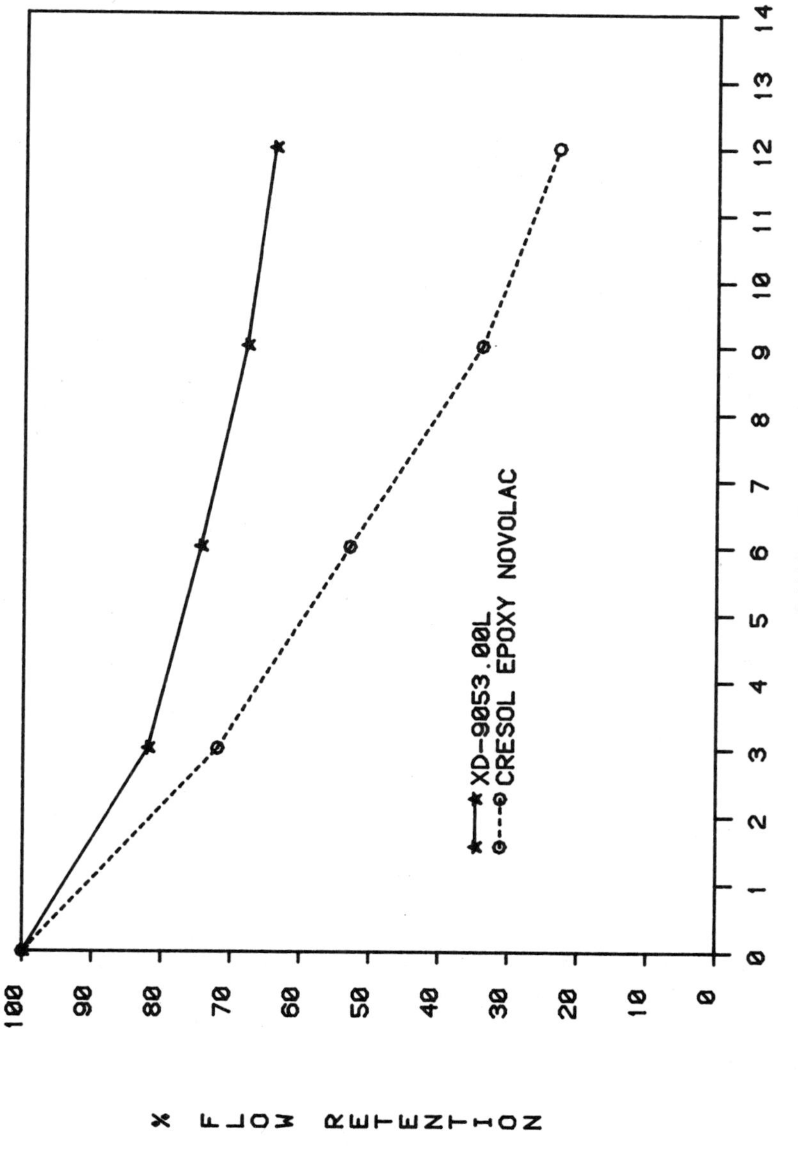

Figure 7. Spiral flow stability XD-9053.00L vs cresol epoxy Novolac accelerated phenolic-cured transfer molding compound.

Extrapolation of this data indicates that the imidazole accelerator level in XD-9053.00L could be increased in order to further quicken cure while maintaining equivalent stability to other more conventional systems.

Effects of Hot-Wet Exposure. In advanced aerospace adhesives and composites the effect of exposure to both hot and wet environments is critical to long-term aircraft performance. Table V shows the results of exposing XD-7342.00L and TGMDA both cured with DADS to 65.5°C heat at 100% relative humidity for 120 days. These conditions assure that the polymer matrix has absorbed an equilibrium amount of moisture which tends to plasticize a resin system and lower physical properties. As the data indicates, XD-7342.00L maintains a higher percentage of its original flexural strength after exposure. XD-7342.00L maintained over 60% of its flexural strength at room temperature and at 180°C after exposure, and the TGMDA system maintained less than 50% strength at both temperatures. The initially higher heat distortion of XD-7342.00L, coupled with its superior hot-wet aging, makes this resin an important candidate for the rapidly-growing aerospace composite industry.

Summary. Even though epoxy resins were introduced to industry over two decades ago, there remains room for innovation and improvement in the chemical design of new epoxy resins. Resins based on the triglycidyl ether of tris(hydroxyphenyl)methane are examples of a new generation of products with superior properties over conventional resins, and they are opening new markets where epoxies will be the material of choice. Two rapidly-growing areas where these products offer utility are in encapsulation of sophisticated electronic devices such as semiconductors and in manufacturing high performance composites and adhesives for the aerospace industry. Experimental Semisolid Epoxy Resin XD-7342.00L fits the needs of aerospace applications because of its high temperature strength retention and improved hot-wet properties. XD-9053.00L is aimed at electronic encapsulation where purity, reactivity, formulated stability and high temperature resistance are important. While other new product modifications based on this technology are being investigated at the early research stage, these two experimental products are currently available in semicommercial quantities.

TABLE V

TYPICAL EFFECT OF LONG-TERM MOISTURE EXPOSURE
ON
FLEXURAL STRENGTH OF CLEAR CASTINGS OF XD-7342.00L
COMPARED WITH TETRAGLYCIDYL METHYLENEDIANILINE
CURED WITH DIAMINODIPHENYL SULFONE*

	XD-7342.00L		TETRAGLYCIDYL METHYLENEDIANILINE	
	BEFORE EXPOSURE	AFTER EXPOSURE	BEFORE EXPOSURE	AFTER EXPOSURE
FLEX. STRENGTH, PSI				
ROOM TEMP.	16,600	11,100	10,900	5,300
180°C	12,900	7,900	10,500	5,183
FLEX. MODULUS, PSI				
ROOM TEMP.	4.1×10^5	-	5.4×10^5	4.7×10^5
180°C	3.2×10^5	2.3×10^5	3.1×10^5	1.9×10^5
% STRENGTH RETENTION		67		49
% WEIGHT CHANGE		4.1		5.6

*SYSTEMS GELLED AT 105°C, POSTCURED 8 HOURS @ 170°C, 12 HOURS @ 200°C, EXPOSED 120 DAYS @ 65.5°C AND 100% RELATIVE HUMIDITY

RECEIVED December 2, 1982

8

Fast Curing Epoxy–Episulfide Resin for Uses at Room Temperature

WENHSIUNG KU and JAMES P. BELL

University of Connecticut, Institute of Materials Science, Storrs, CT 06268

A fast curing epoxy-episulfide resin system has been developed which has important advantages over standard room temperature formulations. Rheovibron and thermomechanical measurements show two transition temperatures, at approximately 75°C and 135°C, when the polyamide curative Versamide 140 is used. The area under the tan δ curve is very broad and indicates good toughness. The heat of polymerization is very low. Water absorption is lower than for a control system without episulfide. The presence of the episulfide homopolymer network in the standard epoxy matrix has a slight lowering effect on the shear strength, but the difference between wet and dry properties is small.

Development of an epoxy-type resin system which will become hard in 5-20 minutes for small quantities at room temperature, without loss of properties relative to standard room temperature epoxy resins, has been a goal of epoxy research for many years (1). Such a resin would appear to have use in a variety of applications near room temperature, from dentistry and orthopedic bone cement, to industrial and electronic assembly operations where high room temperature curing rates and good properties are required.

Our approach has been to synthesize the diepisulfide resin corresponding to the normal bisphenol A type diepoxide resins. A polyamide type curing agent (Versamide 140) was used because of our particular interest in orthopedic adhesives, i.e. "bone cement". Some of the properties were therefore tailored to be optimum near body temperature. We have found very little prior literature on diepisulfide resins. As prepared, the diepisulfide resin analog of DGEBA is a crystalline solid which must be heated above its melting temperature for reaction. Charlesworth (2) has reported mechanical relaxation behavior of epoxy-episulfide poly-

0097–6156/83/0221–0153$06.00/0

mers prepared at elevated temperature. Leu et al. (3,4) have reported in the patent literature that such polymers have an accelerated rate of cure.

Experimental

A standard bisphenol A type epoxy resin, Epon 828 from Shell Chemical Co. U.S.A., and an amine type curing agent, Versamid 140 from General Mills Chemical, U.S.A., were used as received. The diepisulfide resin was synthesized from Epon 828.

Synthesis and Formulation. The synthesis method is based upon the work of V. Calo et al. (5), in which oxiranes were converted into thiiranes:

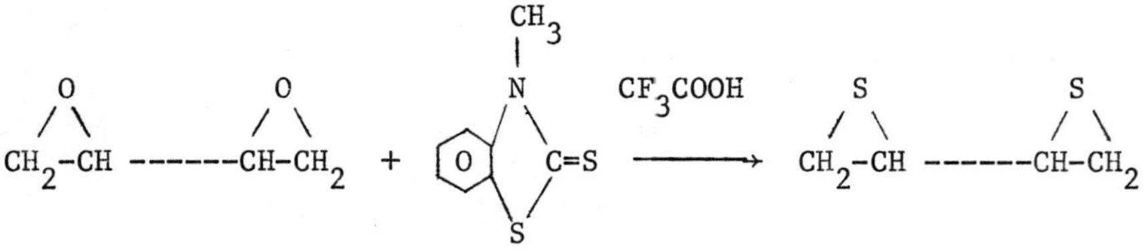

Epon 828 NMBTT DEPPE

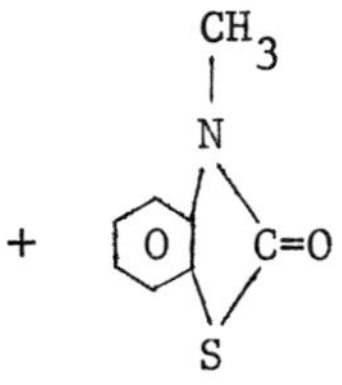

NMBTL

To a round bottom flask containing 13.6g. n-methylbenzothiazol-2-thione (NMBTT), 17 gm $Cf_3CO_2H_3$ and 100 ml. CH_2Cl_2, was added slowly a solution of 14.5 g. Epon 828 in 100 ml. CH_2Cl_2 which had been placed in a 250 ml. separatory funnel. The reaction was carried out with stirring at room temperature (22°C) for thirty minutes. Ground sodium carbonate (20 g.) was then added into the flask and the entire system was stirred for 3 hours. The addition of sodium carbonate was not only to neutralize the CF_3COOH in the reaction mixture, but also to initiate the formation of episulfide product. The sodium trifluoroacetate formed during neutralization was filtered out by means of a Buchner funnel. The filtrate was dried under vacuum to form a very viscous yellowish liquid which was ready for purification.

Silica gel (Silicar CC-7 from Mallinckrodt Chemical, Inc.) was poured into a column (575 mm long and 1 cm diameter) and was tightly packed with the aid of a vibrator. This column was used to isolate the diepisulfide resin. Two grams of viscous liquid obtained as in the above section was diluted by the addition of 10 ml. of solvent mixture of toluene and acetone (100:1 by volume). The diluted sample was then poured into the column. After the sample was completely absorbed by silica gel, the elution solvent (toluene: acetone = 100:1) was run through the column. The solvent effluent was checked at various time intervals by a thin layer chromatographic (TLC) technique. The first eluted product was collected and dried under vacuum. A white powder-like product (DEPPE resin) was obtained with an overall yield of approximately 80%. Infrared and NMR spectra confirmed the product structure, as shown in Figure 1 and 2. For production of larger quantities, the chromatographic purification method was tedious and it was found that the product could be purified by repeated washings of the product with a solution containing 50 parts methanol to one part water.

The product diepisulfide was a crystalline solid. It was found after several trials that it could be maintained in solution for a few hours without crystallization in a mixed epoxy resin system. The system was comprised of 90% Epon 828 , 5% Epon 836 and 5% Epon 1001. These three resins are all of the diglycidyl ether of bisphenol A type, differing only in the number of bisphenol A repeat units in the molecule. By using such a mixture a better cohesive energy density match is obtained, and the broad molecular weight distribution may also inhibit crystallization of the diepisulfide. The final resin system thus contains the liquid diepisulfide-diepoxide mixture, and liquid curing agent unit.

<u>Gelation Time Measurement</u>. A modified method from that described in ASTM D2471 was adopted for the measurement of gelation time for this new resin system. Epoxy-episulfide resin and Versamid 140 were weighed into separate regions of one small aluminum dish and brought to room temperature (23°C). The two resins were not permitted to contact each other in the dish. The two components were then mixed thoroughly for 30 sec. by the use of a spatula. The resulting resin was transferred to a silicon rubber mold which measured 1/4" X 3/16" X 3/16". The rubber mold containing the sample was placed on an insulated cardboard surface in still air at 23°C. The resin was then probed with a wooden applicator stick (6 inches long, from VWR Scientific), holding the stick perpendicular to the resin surface. The gelation time was taken as the point at which the resin no longer adhered to the end of the probe.

<u>Sample Preparation</u>. For an epoxy resin system, a typical method of sample preparation for mechanical testing is as follows: resin

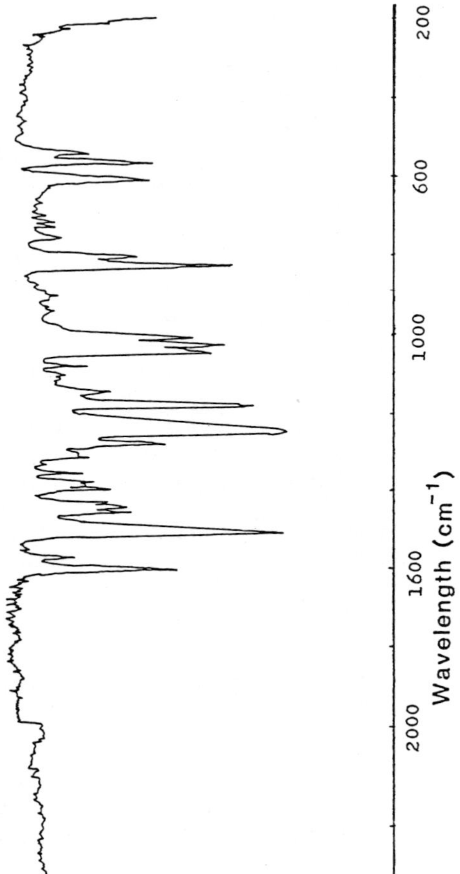

Figure 1. IR spectrum for DEPPE resin.

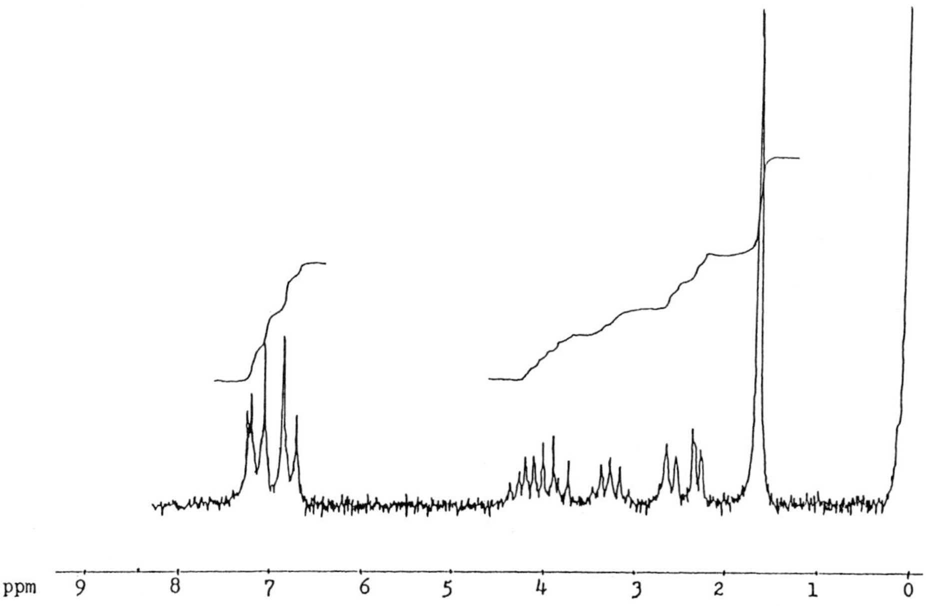

Figure 2. NMR spectrum for DEPPE resin.

and curing agent are mixed, transferred to the mold, vacuumed for about 40 min. to eliminate air in the system, and then cured at a given temperature. It is, however, extremely difficult to obtain good samples for an epoxy-episulfide system by this method because the system cures so rapidly that there is no way to vacuum the air out of the sample before it gels. Since a considerable amount of air is trapped during mixing, one method of air exclusion is to mix the resins and curing agent under vacuum. Figure 3 shows the scheme of a mechanical mixer used in this work. The mixing blades were made of Teflon and were trimmed and bent to match the curvature of the vacuum flask. The epoxy-episulfide resin was weighed into the vacuum flask and the curing agent was stored in the calibrated separatory funnel. Under vacuum, the Teflon stopcock of the separatory funnel was opened to permit the curing agent to be pulled into the vacuum flask, and the resins were mixed by the rotating mixing blades. The mixed sample was then poured into a glass mold and cured in a room maintained at 37°C. All samples were cured for 7 days at 37°C. The wet specimens were prepared by soaking the cured samples in a saline solution (0.9 N NaCl solution) at 37°C. The water absorption was measured periodically until no measurable weight change was observed.

Thermal Characterization. The DuPont 941 Thermal Mechanical Analyzer (TMA), attached to a model 900 thermal analyzer, was used for the measurement of glass transition temperature (Tg). The heating rate was set at 10 °C/min. For wet samples, about 0.05 ml of distilled water was dropped into the TMA sample holder tube before running the experiments, to maintain the sample in a water saturated state during the measurement.

The DuPont 990 Thermal Analyzer with a DuPont DSC cell plug-in module was used for the measurement of heat of reaction. Epoxy-episulfide resin and Versamid 140, which were weighed into an aluminum dish, were mixed by spatula for 20 sec. and quenched with liquid nitrogen at once. At the same time, the DSC cell was cooled down to -50°C with the aid of liquid nitrogen. One piece (about 10-20 mg) of quenched sample was transferred to a preweighed hermetic cup which had already been cooled in liquid nitrogen. The sample was then placed on the sample platform inside the DSC cell and the experiment was performed by programming the temperature from -50°C to 140°C with a heating rate of 5°C/min. The heat of reaction, ΔH, was then calculated by the comparison with the heat of fusion of Indium, which was measured at the same conditions.

Physical Property Evaluation. A Rheovibron (direct reading dynamic viscoelastometer, model RHEO-200, manufactured by Toyo Measuring Instruments Co.) was used to measure the temperature dependence of the complex modulus and damping, tan δ, at 11 Hz. An Oscilloscope OS-46 A/U (Sentinel Electric Inc., Philadelphia,

Figure 3. Photograph of apparatus for mixing small quantities of resins with curing agent under vacuum.

PA) was attached to help the phase adjustment and to monitor the preset of optimum tension.

The shear strength measurement was conducted with an Instron tensile test machine of the constant-rate-of-crosshead movement type. A shear tool described in ASTM D732 was used. The samples were run at a crosshead speed of 0.05 in/min.

Results and Discussion

Evaluation of Epoxy-Episulfide Resin System. A plot of gelation time with respect to the DEPPE content in the sample, with the same numbers of equivalents of Epon 828 and Versamid-140, is shown in Figure 4. Gelation time is dramatically reduced when a small amount of DEPPE is introduced. The curing reaction of a conventional epoxy resin system is governed by the nucleophilic addition of amine to the epoxide group. In the case of the epoxy-episulfide resin system, the reaction of amine with the episulfide ring is also included. Kinetic measurements, which will be discussed in a separate publication, have shown that the anionic polymerization of the DEPPE itself plays an important role in the high curing rate of this system.

Water absorption decreases as the amount of DEPPE in the mixture increases, as shown in Figure 5. Note that the data are for equilibrium water absorption in saline solution not at room temperature, but at 37°C. It seems reasonable that the product formed in the epoxy-episulfide system, which involves many —S— linkages, is less polar than the product from a standard epoxy resin system containing many hydroxyl groups after reaction.

Increasing the DEPPE content increases both the dry and water-saturated glass transition temperatures as measured by Thermomechanical Analyzer, as illustrated in Figure 6. Rheovibron data (Figures 7-10) give a more detailed understanding of the glass transition of the epoxy-episulfide system; a second transition at ∿135°C is present and becomes more significant as a greater portion of DEPPE was introduced into the sample. The two glass transition temperatures correspond to two incompatible polymer structures formed in the amine cured epoxy-episulfide resin: one is a conventional epoxy-amine resin which has a Tg at approximately 75°C; the other seems to be an episulfide homopolymer which has a Tg above 130°C.

Rheovibron data also show that the system potentially has a higher upper use temperature than the Epon 828 - Versamid 140 system, since the elastic modulus, E', drops more slowly. The area under the tan δ curve for this system is greater than that of a corresponding epoxy system. The areas under the damping curves are listed in Table I. The data imply improved energy absorption capability for the new system.

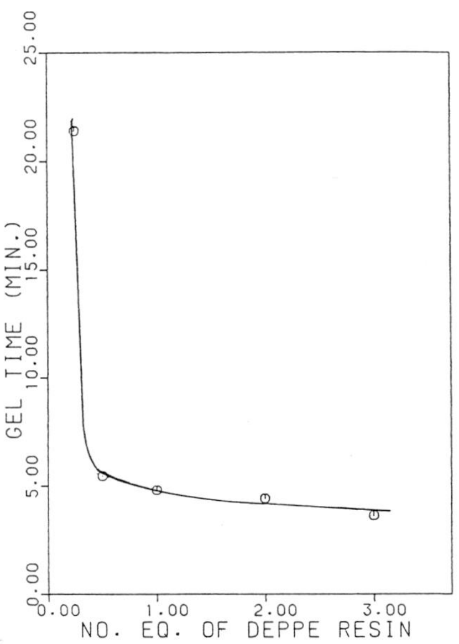

Figure 4. Gel time data for epoxy-episulfide resin system with one equivalent each of Epon 828 and Versamid-140 at 22°C.

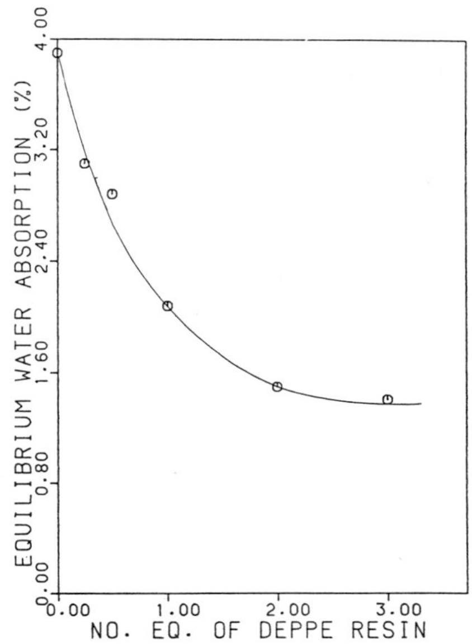

Figure 5. Equilibrium water absorption for epoxy-episulfide resin system with one equivalent of each of Epon 828 and Versamid-140 at 37°C.

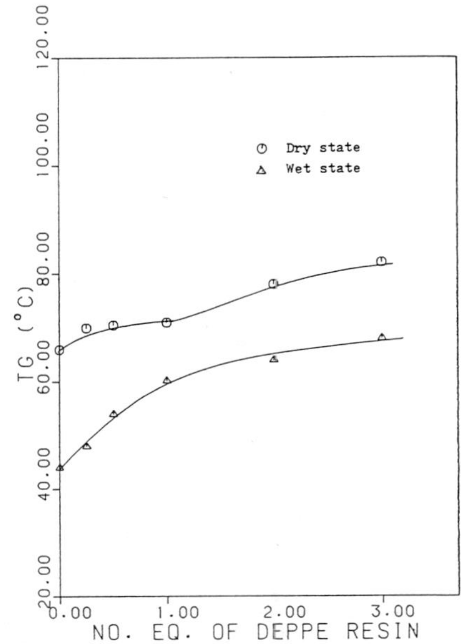

Figure 6. Glass transition temperatures for epoxy-episulfied resin system with one equivalent each of Epon 828 and Versamid-140.

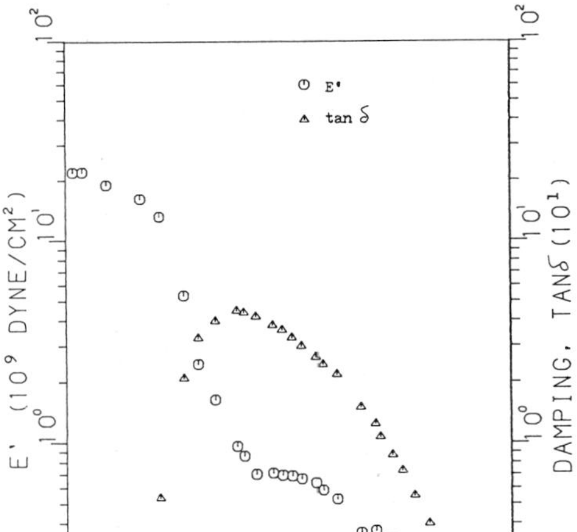

Figure 7. Rheovibron data for #EPS-2 (containing 0.5 equivalent of DEPPE).

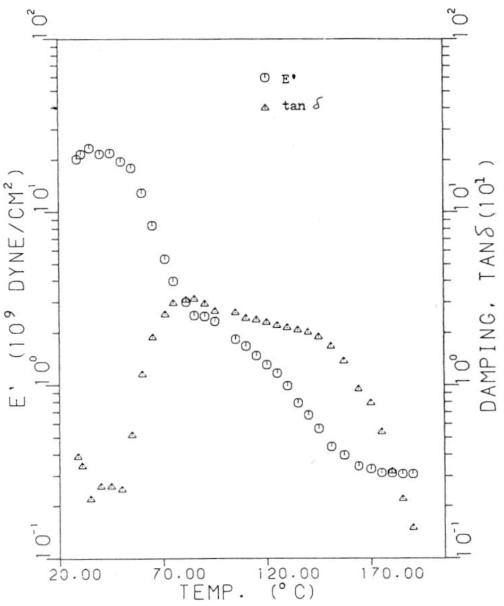

Figure 8. Rheovibron data for sample #EPS-3 (containing 0.75 equivalent of DEPPE).

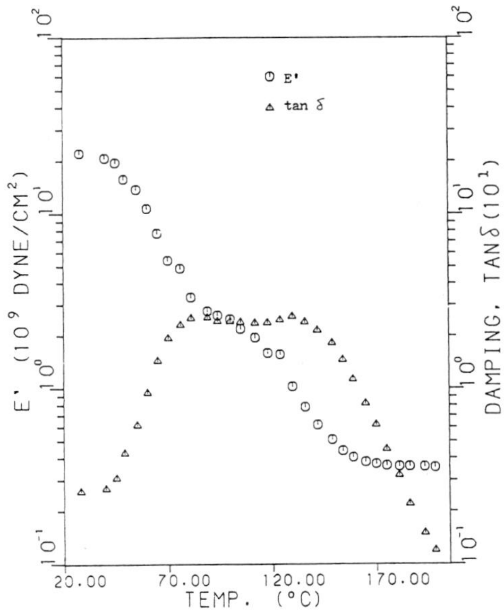

Figure 9. Rheovibron data for sample #EPS-4 (containing 1.0 equivalent of DEPPE).

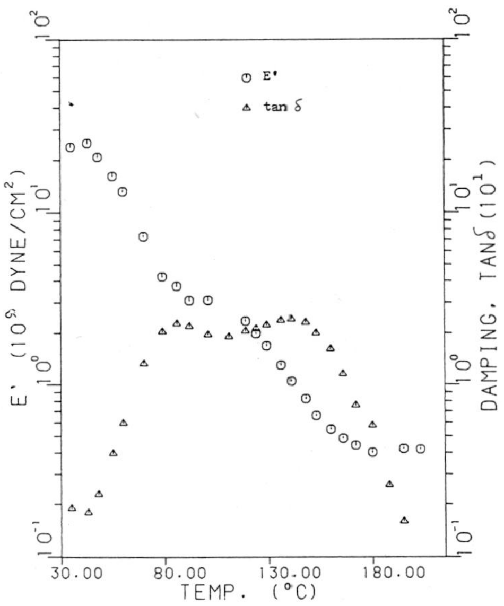

Figure 10. Rheovibron data for sample #EPS-5 (containing 1.5 equivalent of DEPPE).

TABLE I
The area under the damping curve for the epoxy-episulfide resin system

	Equivalence Ratio Epon 828:DEPPE:Versamid 140	Relative Area (in^2)
EPS-1	1.00:0.00:1.00 (Control)	14.99
EPS-2	1.00:0.50:1.00	19.87
EPS-3	1.00:0.75:1.00	20.78
EPS-4	1.00:1.00:1.00	20.19
EPS-5	1.00:1.25:1.00	19.21

The E' values of the samples in the rubbery state, i.e. the E' values at a temperature 40°C above the glass transition, were measured and are summarized in Table II. It is shown that the E' values of the rubbery samples with higher DEPPE content are

TABLE II
The E' value for cured samples in the rubbery plateau region

	Equivalence Ratio Epon 828:DEPPE:Versamid 140	E' (10^9 dyne/cm^2)
AC-1	Simplex, an acrylic cement	----
EPS-1	1.00:0.00:1.00 (Control)	0.169
EPS-2	1.00:0.50:1.00	0.295
EPS-3	1.00:0.75:1.00	0.315
EPS-4	1.00:1.00:1.00	0.368
EPS-5	1.00:1.25:1.00	0.407

* At temperature 40°C above the glass transition temperature.

larger. This means that the crosslinking density increases as the amount of DEPPE increases in the epoxy-episulfide system. The average molecular weight between crosslinking points can be estimated, based on ideal rubber elasticity theory in which

$$G = \rho RT/\bar{M}c$$

where, G is the shear modulus, ρ is the density, R is gas constant, T is absolute temperature and $\bar{M}c$ is average molecular weight between crosslinks. The densities of samples were determined by a density gradient column. They were between 1.10 and 1.20 g/cm^3. When ρ = 1.15 g/cm^3 and G = E'/3 were used, the $\bar{M}c$ results in Table III were obtained. The lower $\bar{M}c$, i.e. higher crosslinking density, is shown for the system having a higher DEPPE content. These results also indirectly confirm the occurrance of highly crosslinked episulfide homopolymer in the sample, since some of the $\bar{M}c$ results are lower than those obtainable from the epoxy polymer alone.

TABLE III
The $\bar{M}c$ value for epoxy-episulfide resin system

Equivalence Ratio Epon 828:DEPPE:Versamid 140	DEPPE (no.eq.)	$\bar{M}c$
1.00:0.00:1.00 (Control)	0	501
1.00:0.50:1.00	0.5	287
1.00:0.75:1.00	0.75	269
1.00:1.00:1.00	1.0	230
1.00:1.25:1.00	1.5	208

The shear strength of the resin as a function of episulfide content is shown in Figure 11, where one observes a greater effect of DEPPE on dry strength than on wet strength. The data show that the shear strength decreases when small amount of DEPPE was introduced, but it increases when more than one equivalent of DEPPE was introduced. This result is possibly explained by the formation of episulfide homopolymer which is not completely compatible with the epoxy chains. At low values of DEPPE we see properties of the epoxy network, and at high values, properties of the DEPPE network. At intermediate mixtures the two networks may not be completely compatible and thus give slightly lower strength than either alone.

Comparison of the new resin system with commercial acrylic bone cement. The Rheovibron data for one acrylic bone cement, Zimmer show a very sharp glass transition and dramatic drop in E' at temperatures above Tg, as indicated in Figure 12. This shows that the acrylic bone cement has a glassy, brittle structure below Tg and is very soft above Tg. The epoxy-episulfide system, however, shows a very broad glass transition which not only has high energy absorption capability, but also can retain high strength within a wide temperature range.

The shear strength of the epoxy-episulfide resin system ranges from 6,400 psi to 7,500 psi, comparable to 6,500 psi for Zimmer and 6,230 psi for Simplex acrylic cements.

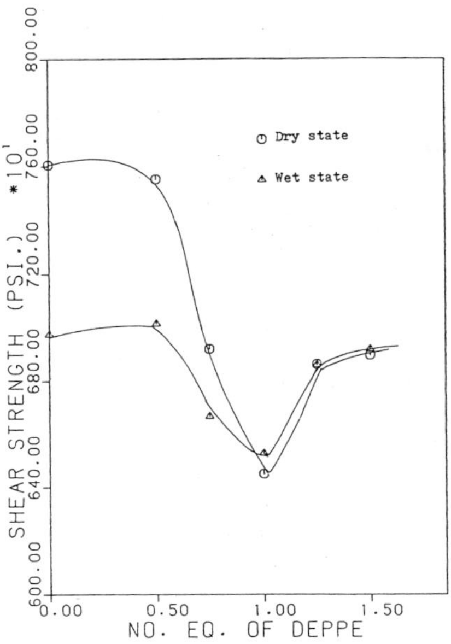

Figure 11. Shear strengths for the epoxy-episulfide resins system with one equivalent each of Epon 828 and Versamid-1409 in the dry and wet states.

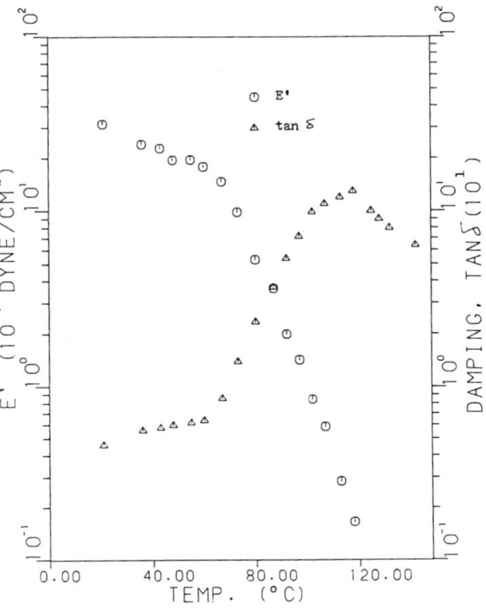

Figure 12. Rheovibron data for bone cement #AC-1.

The heat of reaction for the system (20 mcal/mg) is comparable to the acrylic bone cements (19.3 mcal/mg for Zimmer and 32.4 mcal/mg for Simplex). The reason of the low heat of reaction for the commercial acrylic cements, compared with 136 mcal/mg for the polymerization of MMA monomer (6), is that part of cement has been prepolymerized. Since there is about 10% of inert filler such as $Ba(SO_4)_2$ in the commercial cements, the heat of reaction of the epoxy-episulfide system containing filler would be even lower than the above 20 mcal/mg.

The high room temperature curing rate of the epoxy-episulfide resin system is not only comparable to that of acrylic cement, but also controllable by changing the amounts of the ingredients DEPPE and Versamid 140.

In addition to superior properties of the epoxy-episulfide system, as mentioned above, it does not contain any low molecular weight initiators and/or accelerators such as are present in acrylic bone cement, and the curing reaction is based on a ring opening reaction which would be expected to have lower shrinkage than the free radical polymerization of acrylic cement. Therefore, this new resin system is a tough, low shrinkage material with rapid curing and high mechanical strength. It appears to have many potential uses, both in biomedical and industrial applications.

Conclusions

A Versamid 140 cured epoxy-episulfide resin system, whose gelation time, ranging from several minutes to three hours at room temperature, is controllable by changing the ingredients, has been developed. It has high strength, good toughness, low water absorption and low heat of reaction. Compared with commercial acrylic bone cement, this system shows potential use in several biomedical applications. For some industrial applications and electrical assembly operations, aromatic curing agents can be used to enable the design of high Tg thermosetting materials. The formation of conventional epoxy-amine and also episulfide homopolymer block network structures in the system has been disclosed from Rheovibron data. A chemical investigation of reaction kinetics of this system has been completed and will be published in a separate paper.

Literature Cited

1. Lee, H.; Cupples, A. L.; Stoffey, D. G. "Epoxy Resins", p. 173-207, ADVANCES IN CHEMISTRY SERIES No. 92, ACS: Washington, D.C., 1970.
2. Charlesworth, J. M. J. Polym. Sci. Polymer Phys. Ed., 1979, 17(2), 329.
3. Leu, G.; Sulsen, U.; Widmer, J.; Goeth, J.; Ger. Offen. 2,505,368, Sept. 4, 1975.

4. Leu, G.; Sulsen, U.; Widmer, J.; Goeth, J.; Swiss 598,297, April 28, 1978.
5. Calo, V.; Lopez, L.; Marchese, L.; Pesce, G. J.C.S. Chem. Comm., 1975, 621.
6. Brandrup, J.; Immergut, E. H. Editors, "Polymer Handbook"; VI-69, 2nd printing, John Wiley & Sons, Inc. 1969.

RECEIVED December 16, 1982

9

Influence of Physical Aging on the Time-Dependent Properties of Network Epoxies and Epoxy–Matrix Composites

ERIC S. W. KONG

Stanford University/NASA-Ames Joint Institute for Surface and Microstructure Research, Stanford University, Department of Materials Science and Engineering, Stanford, CA 94305, and

NASA Ames Research Center, Materials Science and Applications Office, Moffett Field, CA 94035

Volume relaxation and enthalpy relaxation processes have been known to occur in amorphous, glassy materials for many years (1). This relaxation phenomenon involves a continuation of vitrification after the material is quenched into the glassy state. Struik, who coined the term "Physical Aging" (2), demonstrated that the chain mobility of macromolecules is not quite zero even though the polymer is stored in the glassy state (3). Specifically, when stored at temperatures between the glass transition temperature, Tg, and the highest secondary transition temperature below Tg, T_β, the material approaches in an asymptotic fashion from a quenched, non-equilibrium state to an aged, equilibrium glassy state. Thus, physical aging is viewed as a recovery process that arises from a thermodynamic drive to attain equilibrium (see Figure 1).

Physical aging has been known to affect the properties of thermoplastics (4,5). In fact, a majority of efforts so far have been focused on the study of this recovery phenomenon in linear macromolecular systems (1-5). The effects of physical aging on the time-dependent properties of crosslinked polymer systems or thermosets have been by and large ignored until recently (6). This paper represents a summary of findings thus far on how physical aging influences the durability of crosslinked epoxies and the carbon-fiber-reinforced epoxy composites (7,8).

Experimental

The epoxy used in this study was Fiberite 934 resin supplied by Fiberite Corporation, Winona, Minnesota. The main constituents of this resin are 63.2% by weight of tetraglycidyl 4,4'-diaminodiphenyl

0097–6156/83/0221–0171$06.25/0

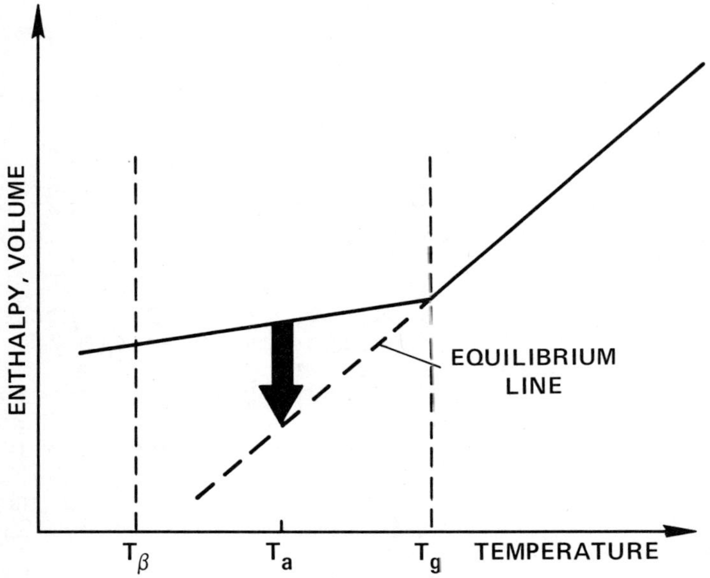

Figure 1. Origin of physical aging as explained from a thermodynamic point of view. T_g is the glass-transition temperature; T_β is the highest secondary transition temperature below T_g; and T_a is the sub-T_g annealing temperature.

methane (TGDDM), 25.3% of the crosslinking agent 4,4'-diaminodiphenyl sulfone (DDS), 11.2% of diglycidyl orthophthalate, and about 0.4% of boron trifluoride/ethylamine catalyst (9).

The neat resin was prepared by casting. The epoxy material was first subjected to degasification at 85°C inside a vacuum oven. The softened liquid was then poured into a preheated mold. The curing schedule was 121°C for 2 h and 177°C for 2.5 h, followed by a slow cooling at ca. 0.5°C per min. to room temperature. Symmetrically reinforced $(\pm 45^{\circ})_{2S}$ Thornel 300/Fiberite 934 laminates were fabricated from prepreg tapes obtained from Fiberite Corporation. The details on specimen fabrication are described elsewhere (8).

With the exception of five specimens (which were to be tested in the as-cast or as-fabricated condition), all specimens were postcured for 16 h at 250 °C, followed by a slow cooling to room temperature at a rate of ca. 0.5°C per min. Testing was performed on the five as-postcured specimens. The other postcured specimens were heated to 260°C for 20 min. and then immediately air-quenched to room temperature inside a desiccator. Five of these as-quenched specimens were tested. The others were sub-Tg annealed in nitrogen atmosphere at 140°C for time increments of 10, 10^2, 10^3, 10^4, and 10^5 min. Time-zero for the sub-Tg annealing experiment was taken as the time when a mercury thermometer placed adjacent to the specimens reached the temperature of 140°C. At each decade of aging time, five specimens were removed from the environmental chamber and stored inside a desiccator at room temperature prior to testing.

To demonstrate the "Thermoreversibility" of physical aging (2), the following requenching procedure was carried out. Some of the specimens aged for 10^4 min. were heated to 260°C, which is above the epoxy Tg, for 20 min. and then air-quenched to room temperature inside a desiccator. Five of these requenched specimens were tested immediately, while the rest were subjected to "re-aging" at 140°C in nitrogen for the time increments of 10, 10^2, 10^3, 10^4, and up to 10^5 min. At least five specimens were tested for each decade of re-aging time.

A Perkin-Elmer DSC-2 differential scanning calorimeter was used to measure the heat capacity of the neat epoxies. The DSC analysis was performed in nitrogen at a heating rate of 10°C per min. The DSC was coupled to a "scanning-auto-zero" unit for baseline optimization. Each specimen was measured from 50°C to 280°C. The specimens were discs of 5 mm in diameter cut from 0.8 mm thick resin-sheet cast using the method described above. The enthalpy relaxation measurements were made by superimposing the first and second scans for each specimen using a data analysis suggested by Wyzgoski (10).

An Instron model 1122 was used for the stress relaxation experiments. Dog-bone-shaped specimens were prepared in accordance to ASTM D1708-66. The specimens were 22.25 mm long (linear section of the dog-bone-shaped specimen), 4.75 mm wide, and ca. 1.5 mm thick. The resin components were prepared as described above and poured into a preheated Dow Corning silicon rubber RTV 3110 mold. In the stress relaxation experiment, the specimens were stretched within a 2-sec. interval to an elongation of ca. 1%. The stress level was then monitored as a function of time under constant elongation. The results are given as the percent of stress relaxation in the first 10 min. of the experiment. At least five specimens were tested for each decade of sub-Tg annealing time.

Dynamic mechanical analysis was carried out using a low-frequency, forced-oscillation DMA 981 unit (E. I. du Pont de Nemours and Company) interfaced with a MINC 11 computer (Digital Corporation, Marlboro, Massachusetts). Measurements were made in nitrogen atmosphere from -100°C to 300°C at a heating rate of 5°C per min. The specimens for the dynamic mechanical experiment were rectangular $(\pm45^{\circ})_{2S}$ laminates with a length of 23.8 mm, width 12.7 mm and typical thickness of 1.0 mm.

Thermal mechanical analysis was performed using a Perkin-Elmer TMS-2 unit. Thermal expansion behavior was monitored in helium atmosphere from 50°C to 260°C at a heating rate of 5°C per min. The specimens were discs of 6 mm in diameter cut from a 2.5 mm thick resin-plate.

Results and Discussion

In an earlier report (8), the effect of sub-Tg annealing on the mechanical properties of carbon-fiber-reinforced epoxy laminates was discussed. In short, it was found that the ultimate-tensile-strength, strain-to-break, and static toughness of $(\pm45^{\circ})_{4S}$ symmetrically reinforced TGDDM-DDS epoxy composites decreased with incresing sub-Tg annealing time at 140°C. In the present study, the kinetics of aging have been followed at 140°C in nitrogen atmosphere using a variety of techniques. It is interesting to note that no weight loss was observed in these materials during the sub-Tg annealing experiments. In this report, results from four different techniques will be discussed, namely, differential scanning calorimetry, stress relaxation, dynamic mechanical analysis, and thermal mechanical analysis.

Differential Scanning Calorimetry. The thermal analysis technique of differential scanning calorimetry has been amply demonstrated in the past to be a useful method to follow the kinetics of enthalpy relaxation in polymers (4,6,10,11). Figure 2 shows the DSC scans of the fully-crosslinked epoxy specimens which were quenched from above Tg and then subjected to aging at 140°C. The full line is the first scan, and the dotted line represents the second scan taken right after rapid cooling from the initial one. The following observations were made:

1. The enthalpy relaxation peak appears near the onset of the transition from the glassy state to the rubbery state. This peak appears only after 10 min. of aging at 140°C.

2. During sub-Tg annealing, the relaxation peak has both shifted to a higher temperature and grown in magnitude (see Figure 2).

3. This recovery phenomenon is thermoreversible. Upon re-aging the material which was cooled from above Tg, the relaxation peak will reappear and grow with time (see Figure 3).

As mentioned earlier, it is possible to measure the relaxation enthalpy using procedures suggested in the literature (6,10). Figure 4 shows the enthalpy loss during the physical aging versus logarithmic sub-Tg annealing time at 140°C. It is apparent that there is a linear relationship between this enthalpy relaxation process and the logarithmic aging time. In an earlier report (8), a linear relationship was also observed between the drop in ultimate tensile behavior and the logarithmic aging time in a similar TGDDM-DDS epoxy system.

Stress Relaxation Experiments. Figure 5 shows the stress relaxation curves of the fully-crosslinked epoxies with 10, 10^2, 10^3, 10^4, and 10^5min. of sub-Tg annealing at 140°C. As thermal aging progresses, the initial stress level increases and the relaxation rate decreases. The higher initial stress is caused primarily by the higher modulus (as also confirmed by dynamic mechanical analysis)but is also partially due to less relaxation during the stretching period. The relaxation rate is measured as a percent of stress relaxation during the first 10 min. of the experiment.

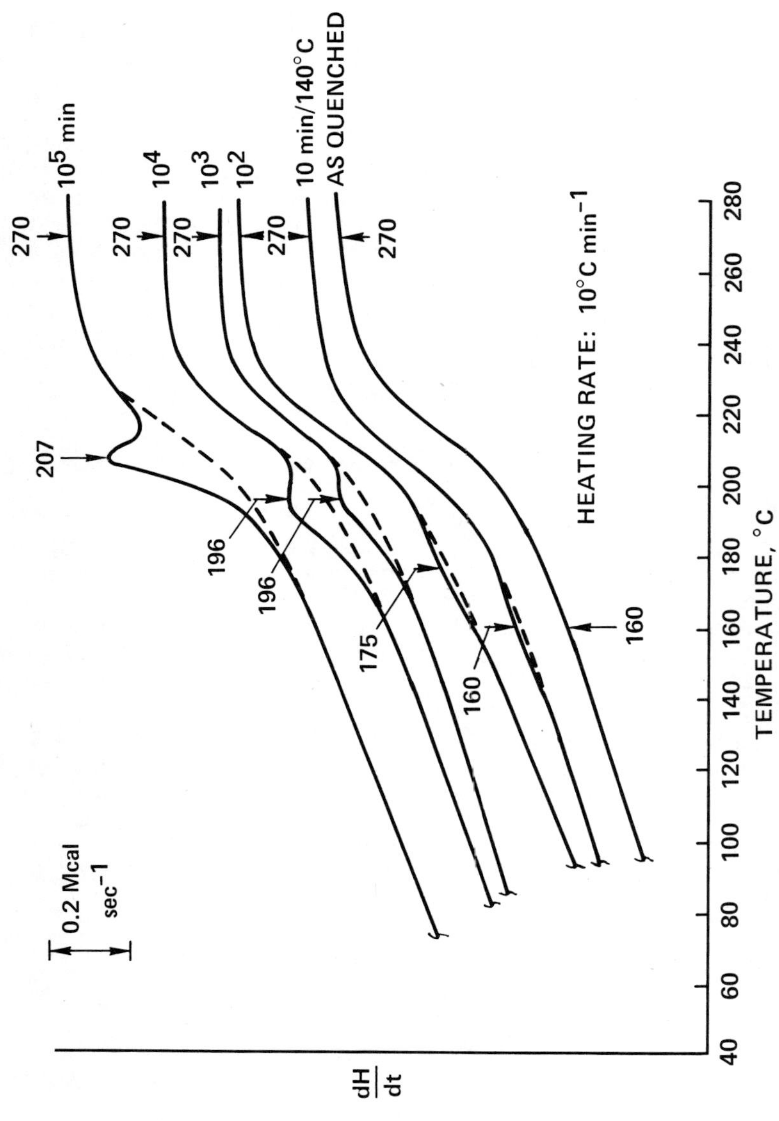

Figure 2. Sub-T_g annealing effects on the DSC traces of quenched Fiberite 934 epoxy resins.

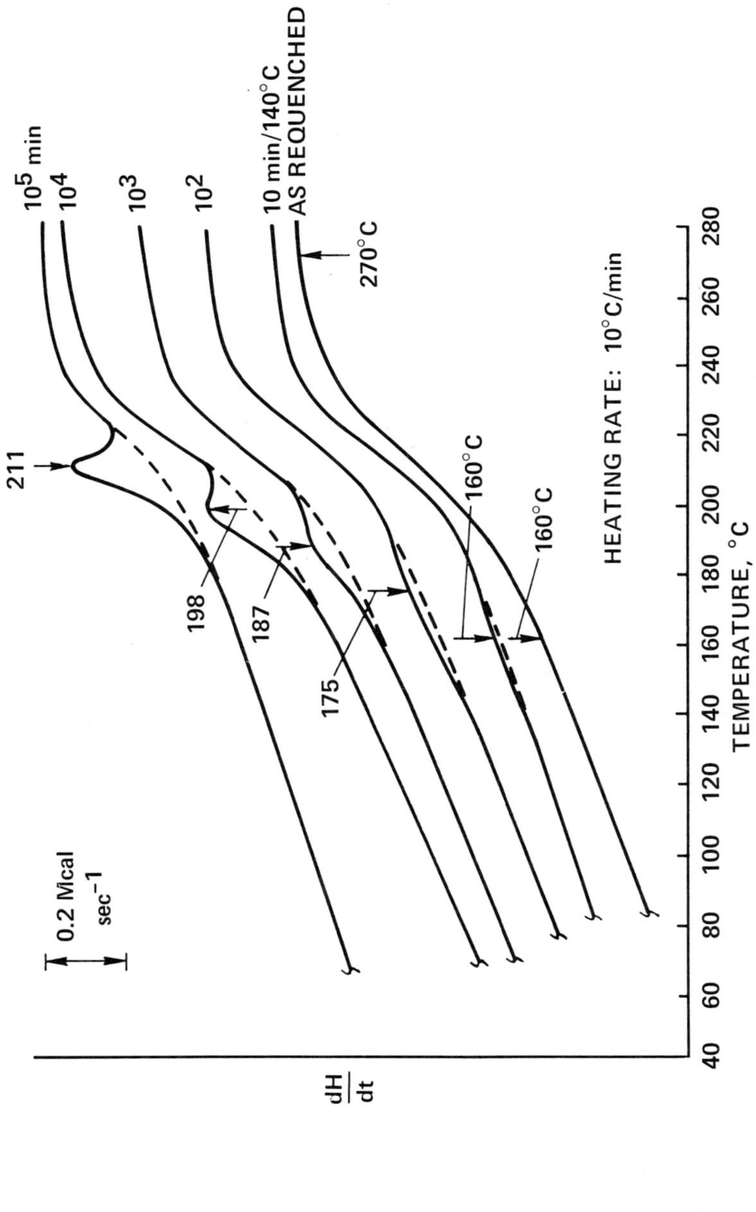

Figure 3. Sub-T_g annealing time effects on the DSC traces of requenched Fiberite 934 resins.

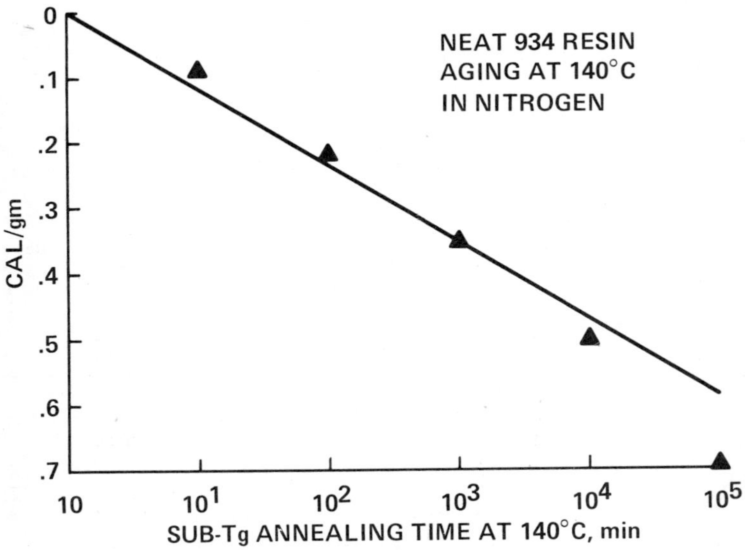

Figure 4. Enthalpy loss during physical aging.

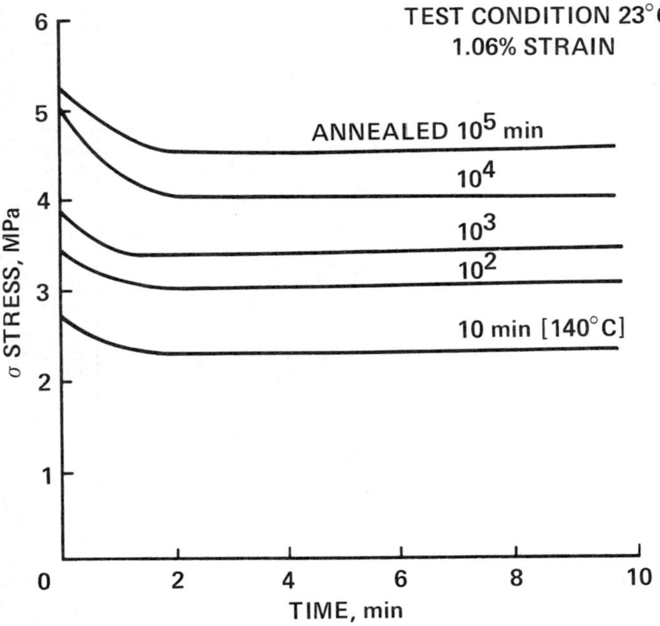

Figure 5. Stress relaxation curves of Fiberite 934 epoxy resins at 23°C.

Figure 6 shows this percent of stress relaxation as a function of logarithmic aging time for the epoxy resin. The data clearly demonstrate that the rate of stress relaxation decreases with aging time. Again, the thermoreversible nature of physical aging is demonstrated by similar data obtained for re-aged specimens (see Figure 7). Samples were aged at 140°C for 10^4 min., annealed above Tg (a "memory-wiping-out process in which the aging thermal history were erased), quenched, and then subjected to aging for the second time at 140°C in nitrogen. The percent of stress relaxation as a function of logarithmic re-aging time for the epoxy resin is shown in Figure 7.

This time-dependent behavior would be expected based on the model that as the network chains lose mobility during the aging process (due to the decrease in free volume), the ability to dissipate stress is reduced. The decrease in stress relaxation rate can therefore be explained on the basis of free volume collapse during physical aging.

Dynamic Mechanical Analysis. The effect of postcuring, quenching, and sub-Tg annealing on the mechanical dispersion peaks of $(\pm 45^{\circ})_{2S}$ carbon-fiber-reinforced epoxy composites is shown in Figures 8 and 9. Figure 8 shows specifically the effect of postcuring and quenching on the damping behavior of the β-transition which occurs in the -55°C region. The low-temperature β-dispersion peak indicates a decrease in damping after postcuring. The postcuring procedure has resulted in an epoxy matrix which has a higher crosslinking density and a lesser amount of free volume. However, after a brief anneal above Tg followed by a quench, the damping of the β-transition increases with respect to the postcured specimens. The increase in mechanical damping can be explained on the basis of increased free volume upon quenching.

Figure 9 shows the effect of physical aging on the decrease in damping as a function of sub-Tg annealing time. This gradual decrease in the magnitude of the secondary dispersion peak can be explained by the relaxation model in which the epoxy network loses mobility as free volume decreases during its asymtotic approach towards the equilibrium glassy state, and, as a result, the ability to dissipate energy is reduced.

This is a significant observation in view of the fact that the area under the secondary mechanical dispersion peak is often correlated to the impact strength of the material (12-14). Because of the absence of bulky side-groups in the thermoset system

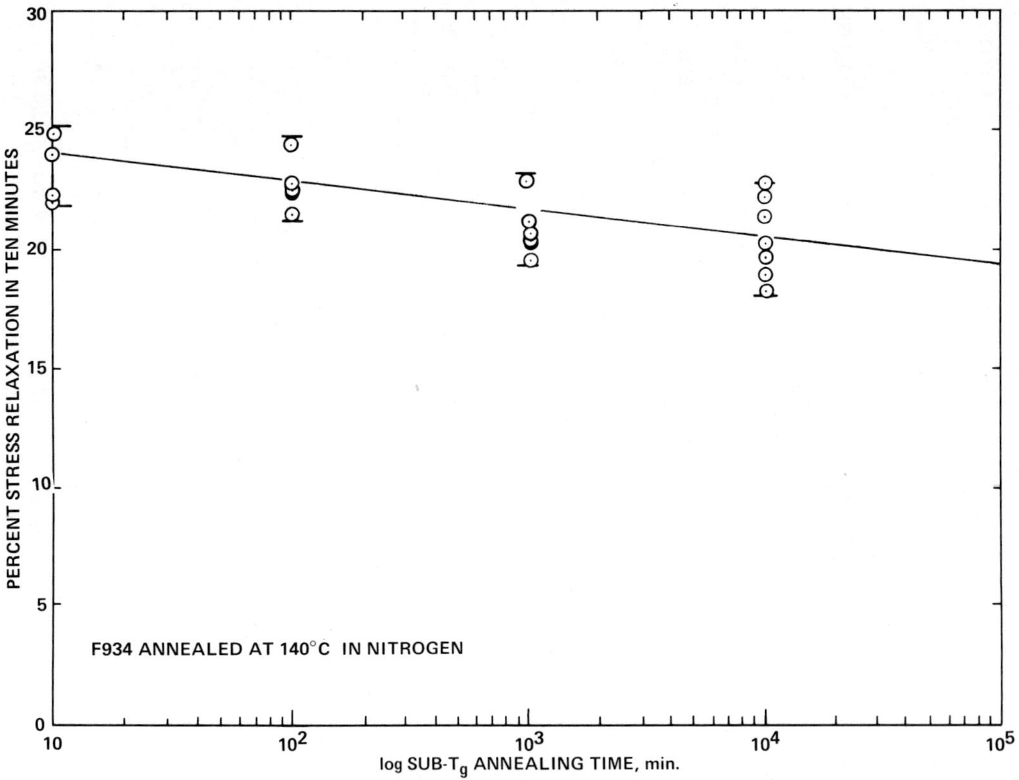

Figure 6. Percent stress relaxation of quenched Fiberite 934 epoxy resins as a function of sub-T_g annealing time.

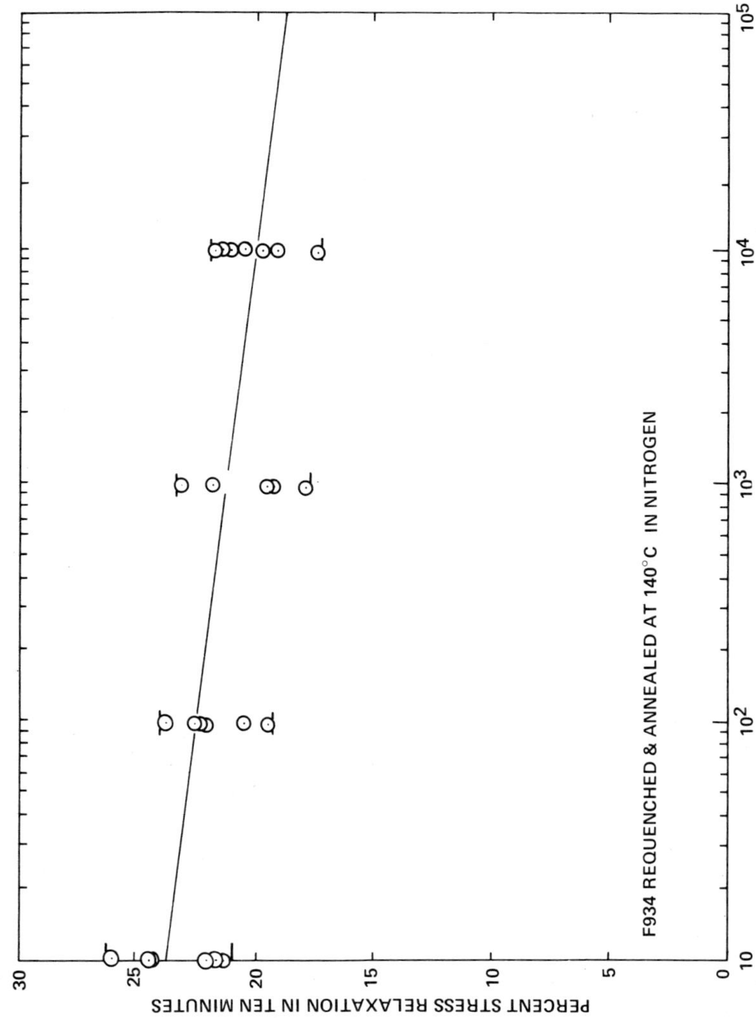

Figure 7. Percent stress relaxation of <u>requenched</u> Fiberite 934 epoxy resins as a function of <u>sub-T_g</u> annealing time.

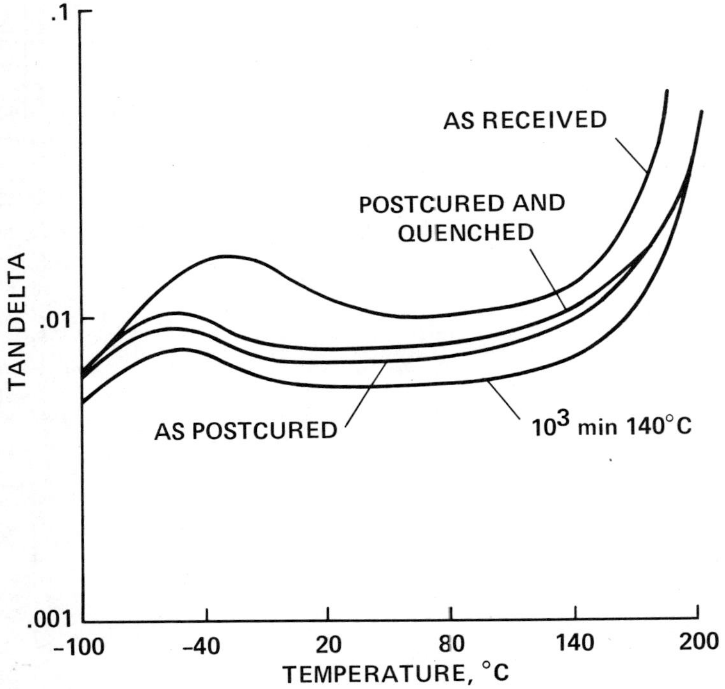

Figure 8. Loss factor (tan <u>delta</u>) for quenched Thornel 300/Fiberite 934 composite specimens as a function of thermal history.

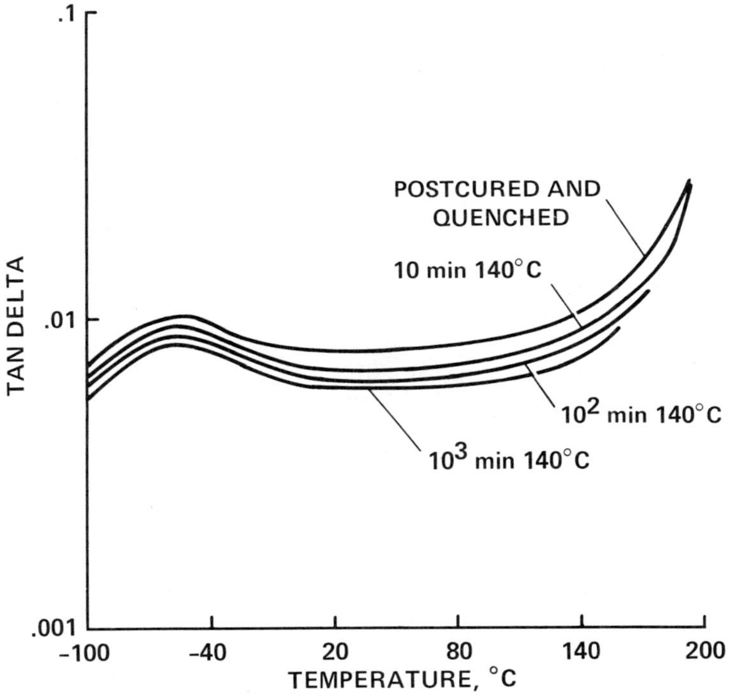

Figure 9. Loss factor for quenched Thornel 300/ Fiberite 934 composite specimens as a function of sub-T_g annealing time.

in this study, it is likely that the origin of the secondary transition can be traced to the motion of the main backbone (13,14). Upon requenching from above Tg and re-aging such material, the thermoreversible nature of physical aging is again demonstrated. The effect of aging on the β-transition in the epoxy matrix of requenched specimens is shown in Figure 10.

Thermal Mechanical Analysis. Thermal expansion experiments served to measure both Tg and volume changes caused by physical aging. The glass transition temperature can be characterized by a slope change in the thermal expansivity of the resin (see Figure 11). As-cast resins show a Tg of 193°C. With postcuring and the resulted increase in chemical crosslinking, the resins show an increase of Tg to 196°C. Quenching of the postcured material resulted in a resin with a even higher Tg of about 214°C.

Figure 12 shows the thermal expansion behavior of well-cured Fiberite 934 epoxy as a function of sub-Tg annealing time at 140°C. The full line represents the first scan while the dotted line represents the second scan after a rapid cooling from the initial scan. The second scan therefore represents the thermal expansion behavior of an as-quenched resin. By superimposing the two curves at the higher temperature section (above Tg), in which the material is at equilibrium in the rubbery state, it can be seen that the aged specimen had a lesser volume than the as-quenched. When the aged specimen was heated, its volume recovered near Tg and approached that of the quenched specimen. It is obvious from the data in Figure 12 that the longer the aging time, the larger amount of volume was lost during sub-Tg annealing. The main contribution of the expansivity experiments is to directly demonstrate that densification is associated with the physical aging process in network epoxies.

Figure 13 shows the thermal expansion behavior of the epoxy as a function of re-aging time at 140°C. Again, the thermoreversibility of physical aging was demonstrated.

Conclusion

Time-dependent changes in the physical properties of a network epoxy and its carbon-fiber-reinforced composite have been observed and explained on the basis of physical aging. In view of the increasing

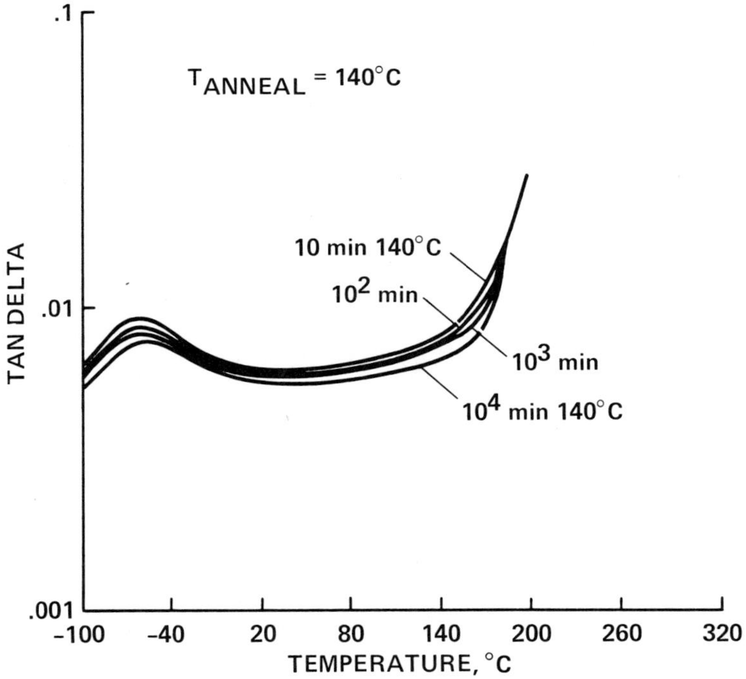

Figure 10. Loss factor for <u>requenched</u> Thornel 300/ Fiberite 934 composite specimens as a function of sub-T_g annealing time.

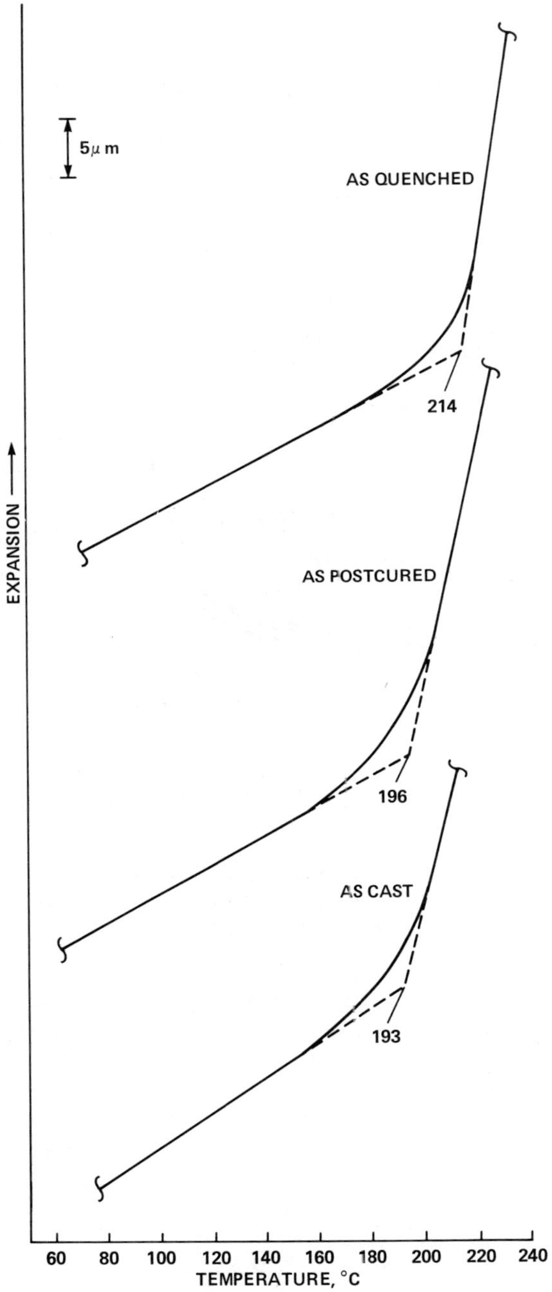

Figure 11. Thermal expansion behavior of Fiberite 934 epoxy resins as a function of thermal history.

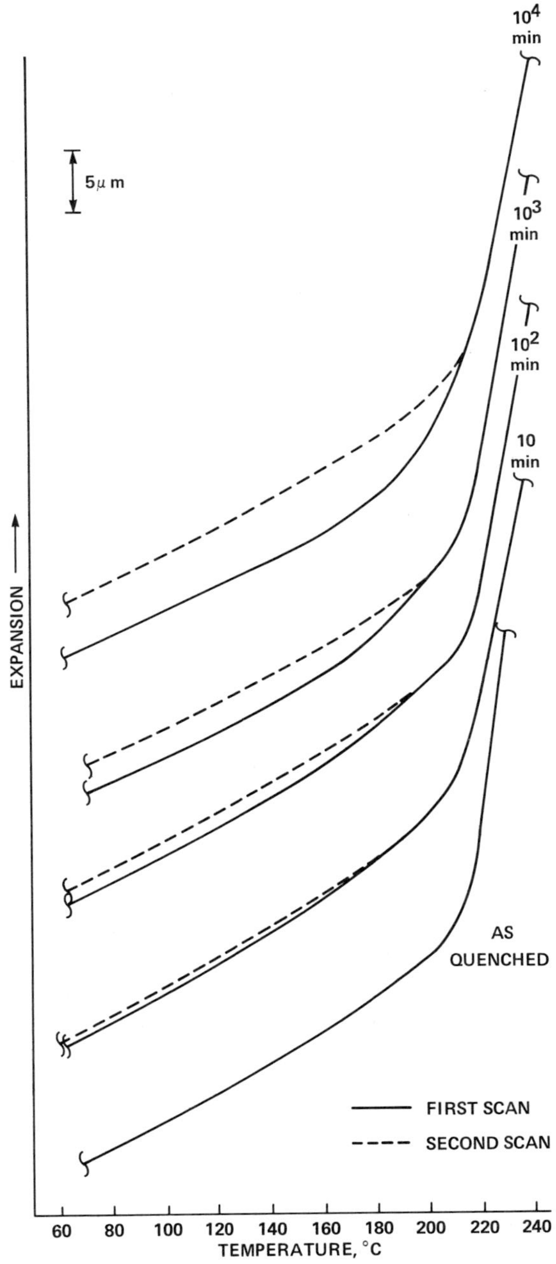

Figure 12. Thermal expansion behavior of fully-crosslinked epoxy resins as a function of sub-T_g annealing time at 140°C.

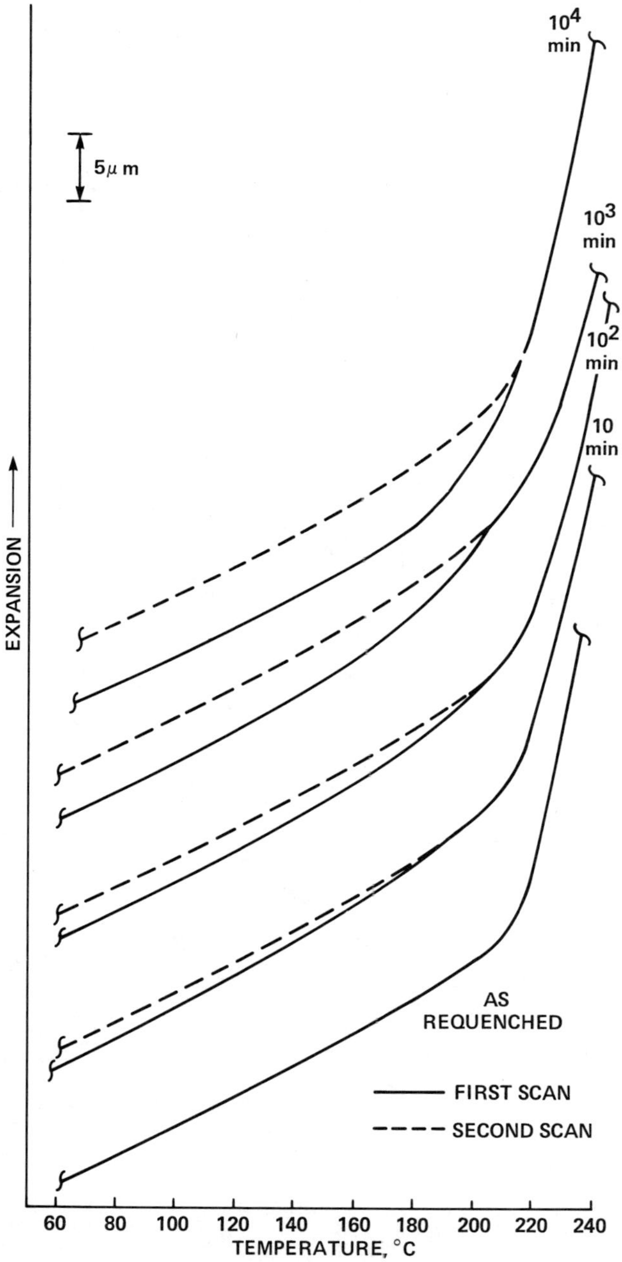

Figure 13. Thermal expansion behavior of fully-crosslinked Fiberite 934 epoxy resins as a function of reaging time.

usage of such materials as structural components in the aerospace and automotive industry (15), parameters such as physical aging time and temperature must be taken into consideration in the prediction of long-term durability and reliability of polymer-matrix composites.

The results from this investigation can be summarized as follows:

- Enthalpy relaxation has been observed in TGDDM-DDS network epoxies.
- Thermoreversible nature of enthalpy relaxation has been demonstrated in the network glasses of TGDDM-DDS epoxy.
- The rate of stress relaxation in neat TGDDM-DDS thermoset resin decreases with sub-Tg annealing time.
- The secondary mechanical damping of epoxy matrix in Thornel 300/Fiberite 934 composite decreases with physical aging time.
- Thermoreversibility of physical aging has been demonstrated in the time-dependent rate of stress relaxation as well as in the damping behavior of the β-transition.
- Volume relaxation has been observed in Fiberite 934 epoxies by monitoring the thermal expansion behavior.
- Thermoreversible nature of volume relaxation has been observed in TGDDM-DDS epoxies.

Synopsis

Matrix-dominated properties of a network epoxy and its carbon-fiber-reinforced composite have been found to be affected by sub-Tg annealing in nitrogen atmosphere. Postcured symmetrically reinforced specimens of Thornel 300 carbon-fiber/Fiberite 934 epoxy composite as well as neat Fiberite 934 epoxy resin were quenched from above Tg and given a sub-Tg annealing at 140°C for times up to 2 months (10^5 min.). The damping behavior, stress-relaxation rate, and the enthalpic state of the matrix material were found to decrease as functions of sub-Tg annealing time. Thermal expansion behavior of the resin was also observed to be affected by aging. No weight loss was observed during sub-Tg annealing. The time-dependent change in physical properties is explained on the basis of free-volume collapse that is related to the volume recovery process of the non-equilibrium glassy network of epoxy. Physical aging in network epoxy has been demonstrated to be "thermoreversible". The aging kinetics was followed by differential scanning calorimetry, stress-relaxation tests, dynamic mechanical analysis, and thermal mechanical analysis.

Acknowledgments

The author would like to thank his co-workers in generating part of the data reported in this paper. Thanks are due to Susanna Lee, John Lai and Mark Rosenberg for their technical support. The author also thank Michael Adamson, Linda Clements, Theodore Sumsion and Howard Nelson for their valuable comments and constructive reviews. A grant (NCC 2-103) from NASA to Stanford University is gratefully acknowledged.

Literature Cited

1. Kovacs, A.J. J. Polym. Sci., 1958, 30, 131.
2. Struik, L.C.E. "Physical Aging in Amorphous Polymers and Other Materials"; Elsevier Scientific Publishing Company: Amsterdam, 1978.
3. Struik, L.C.E. Annals New York Acad. Sci., 1976, 279, 78.
4. Petrie, S.E.B. "Polymeric Materials: Relationships between Structure and Mechanical Behavior"; Amer. Society for Metals: Metals Park, Ohio, 1975, p. 55.
5. Golden, J.H.; Hammant, B.L.; Hazell, E.A. J. Appl. Polym. Sci., 1967, 11, 1571.
6. Ophir, Z.H.; Emerson, J.A.; Wilkes, G.L. J. Appl. Phys., 1978, 49, 5032.
7. Kong, E.S.W.; Wilkes, G.L.; McGrath, J.E.; Banthia, A.K.; Mohajer, Y.; Tant, M.R. Polym. Eng. Sci., 1981, 21, 943.
8. Kong, E.S.W. J. Appl. Phys., 1981, 52, 5921.
9. May, C.A.; Fritzen, J.S.; Whearty, D.K. "Exploratory Development of Chemical Quality Assurance and Composition of Epoxy Formulations"; Lockheed Missiles and Space Company: Sunnyvale, California, Air Force Technical Report: AFML-TR-76-112 (1976).
10. Wyzgoski, M.G. J. Appl. Polym. Sci., 1980, 25, 1455.
11. Matsuoka, S.; Bair, H.E. J. Appl. Phys., 1977, 48, 4058.
12. Wyzgoski, M.G. J. Appl. Polym. Sci., 1980, 25, 1443.
13. Meier, D.J., Ed. "Molecular Basis of Transitions and Relaxations" Gordon and Breach Science

Publishers: London, Midland Macromolecular Monographs, Vol. 4 (1978).
14. Bailey, R.T.; North, A.M.; Pethrick, R.A. "Molecular Motion in High Polymers" Oxford University Press: Oxford, 1981.
15. Beardmore, P.; Harwood, J.J.; Kinsman, K.R.; Robertson, R.E. Science, 1980, 208, 833.

RECEIVED December 2, 1982

10

Effects of Impurities on Hydrolytic Stability and Curing Behavior

GARY L. HAGNAUER and PETER J. PEARCE

Army Materials and Mechanics Research Center, Polymer Research Division, Watertown, MA 02172

Preparative liquid chromatography procedures are described for isolating highly pure N,N'-tetraglycidyl methylene dianiline (TGMDA) from commercial TGMDA resins. Synthesis by-products or impurities are found to have a significant effect on the physical properties and curing behavior of the commercial resins. For example, the viscosity at 50°C of purified TGMDA is about 1/10th that of the commercial resins and impurities in the commercial resins accelerate the rate of hydrolysis of TGMDA. The presence of impurities lowers the temperature for the thermal polymerization of TGMDA but raises the glass transition temperature of the thermal polymerization products. However, impurities lower both the curing temperature for the reaction of TGMDA-diaminodiphenyl sulfone mixtures and the glass transition temperature of the reaction products. The heats of reaction are inversely proportional to the epoxy equivalent weights of the TGMDA resins.

Epoxy resins containing N,N'-tetraglycidyl methylene dianiline (TGMDA) are widely used in the manufacture of fiber-reinforced structural composites for aerospace applications. TGMDA is the principle component (ca. 60-80%) in Ciba-Geigy Corporation's Araldite MY720 and in FIC Corporation's Glyamine G-120. Other components in these resins may include chlorohydrins, glycols, dimers, trimers and higher oligomers. Additionally, water, organic solvents and inorganic salts may be present as impurities. It is generally believed that hydrolyzable chlorides, free hydroxyls, and trace impurities have a significant effect on the reactivity of the resin and may adversely influence the reliability and performance of the composite material. However, the effects of impurities on the properties and curing behavior of TGMDA resins have not been investigated systematically and the properties of "pure" TGMDA are unknown.

In this paper, preparative liquid chromatography procedures are described to isolate highly pure TGMDA from the commercial resins, and the properties of purified TGMDA are compared with those of recently manufactured MY720 and G-120 resins. High performance liquid chromatography (HPLC), differential scanning calorimetry (DSC), and DC resistance techniques are applied to evaluate the hydrolytic stability and curing behavior of the resins and resin formulations containing diaminodiphenyl sulfone (DDS).

Experimental

The TGMDA resins Araldite MY720 (batch No. 5093) and Glyamine G-120 (Lot No. 1003) were obtained recently from Ciba-Geigy Corp. and FIC Corp., respectively, and were stored in closed containers at -13°C. The purification was conducted in two stages using a Waters Associates Prep LC/System 500 with Prep PAK-500/Silica columns.

Stage 1. Separation from commercial resin.

Mobile Phase: Methylene Chloride
Flow Rate: 60 ml/min
Concentration: 20% w/v
Inject: 600 ml
Run Time: 40 min

Stage 2. Further Purification of Stage 1 Material.

Mobile Phase: 60% Ethyl Acetate/40% Hexane v/v
Flow Rate: 200 ml/min
Concentration: 20% w/v
Inject: 5 ml
Run Time: 15 min

A Waters Associates ALC/GPC-244 instrument with M6000A solvent delivery systems, M720 system controller, 710B WISP auto-injection system, M440 UV detector, and M730 data module was used for the HPLC analyses. The following conditions were used for reverse phase HPLC:

Column: Radial PAK C_{18} with RCM100 module
Concentration: 0.5% w/v
Inject Volume: 10 μl
Flow Rate: 3 ml/min
Detector: UV 254 nm
Mobile Phase: (65% H_2O/35% CH_3CN) to (50% H_2O/50% CH_3CN), 20 min, grad 6.

(50% H_2O/50% CH_3CN) to (10% H_2O/90% CH_3CN), 10 min, grad 6.

(10% H_2O/90% CH_3CN) to (10% H_2O/20% THF/70% CH_3CN), 10 min, grad 7.

and for size exclusion HPLC:

Column: μ Styragel (1000, 2x500, 3x100 A^{o})
Concentration: 0.5% w/v
Inject Volume: 10μl
Flow Rate: 2 ml/min
Mobile Phase: THF
Detector: UV 254 nm

Reagent grade water was prepared from distilled water using a Millipore Milli-Q2 water purification system. Distilled-in-glass, UV grade tetrahydrofuran (THF) and acetonitrile (CH_3CN) was used as received from Burdick & Jackson Labs. Solutions were prepared in volumetric flasks and were filtered through 0.2μM Millipore membrane filters.

In addition to HPLC, elemental analysis, vapor phase osmometry, Fourier Transform Infrared Spectroscopy (FTIR), GC/MS, H^1 and C^{13} NMR techniques were applied to verify the purity of the resin samples. Epoxy equivalent weights (EEW) were determined by the standard nonaqueous titration method (1) using chloroform as the solvent. A correction for tertiary amine was obtained by conducting the titration in the absence of the quaternary halide.

Viscosity measurements were made at 50°C on a Rheometrics Mechanical Spectrometer Model RMS-7200 in steady shear mode with the cone and plate geometry.

To evaluate the hydrolytic stability of the resin samples, thin layers of the liquid resins were aged at 60°C and 96% relative humidity. The resins were sampled periodically and HPLC was used to monitor the formation of hydrolysis products as a function of aging time. The HPLC operating parameters were similar to those described previously for reverse phase analysis except for the mobile phase conditions which are shown below:

(70% H_2O/30% THF), 25 min.
(70% H_2O/30% THF) to (10% H_2O/30% THF/60% CH_3CN), 20 min, grad. 7
(10% H_2O/30% THF/60% CH_3CN), 5 min.

Curing studies were performed using a DuPont 990 Thermal Analyzer for differential scanning calorimetry (DSC) and a specially designed apparatus for DC resistance measurements. Glass transition temperatures T_g of the polymerization products were determined by DSC analysis. Puriss grade 4,4'-diaminodiphenyl sulfone (DDS) was used as received from Koch-Light Labs.

Results and Discussion

The size exclusion and reverse phase HPLC chromatograms of MY720 and G-120 are shown in Figures 1 and 2. UV-absorbing components are characterized by having peak retention times which depend upon their molecular structure and solubility properties. The size or amplitude of the peaks are directly proportional to the concentrations of the components being monitored; and more than one component may have the same retention time, especially in the case of size exclusion HPLC. In Figure 1, higher oligomers, trimers and dimers of TGMDA have shorter retention times than TGMDA (2163 seconds), which is shown to be the major component in both samples. Better resolution is obtained by reverse phase HPLC. In Figure 2, high molecular weight,

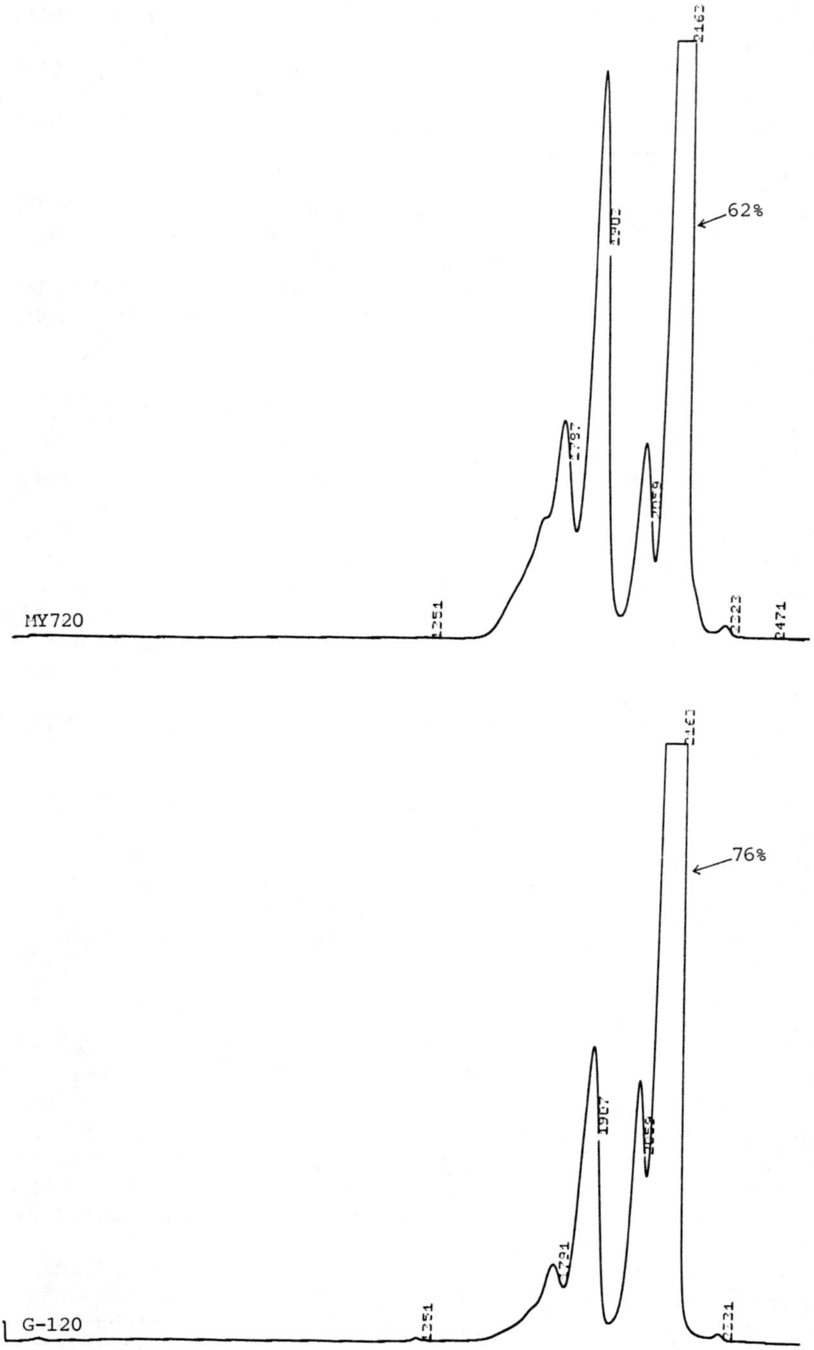

Figure 1. Size Exclusion HPLC of Commercial TGMDA Resins.

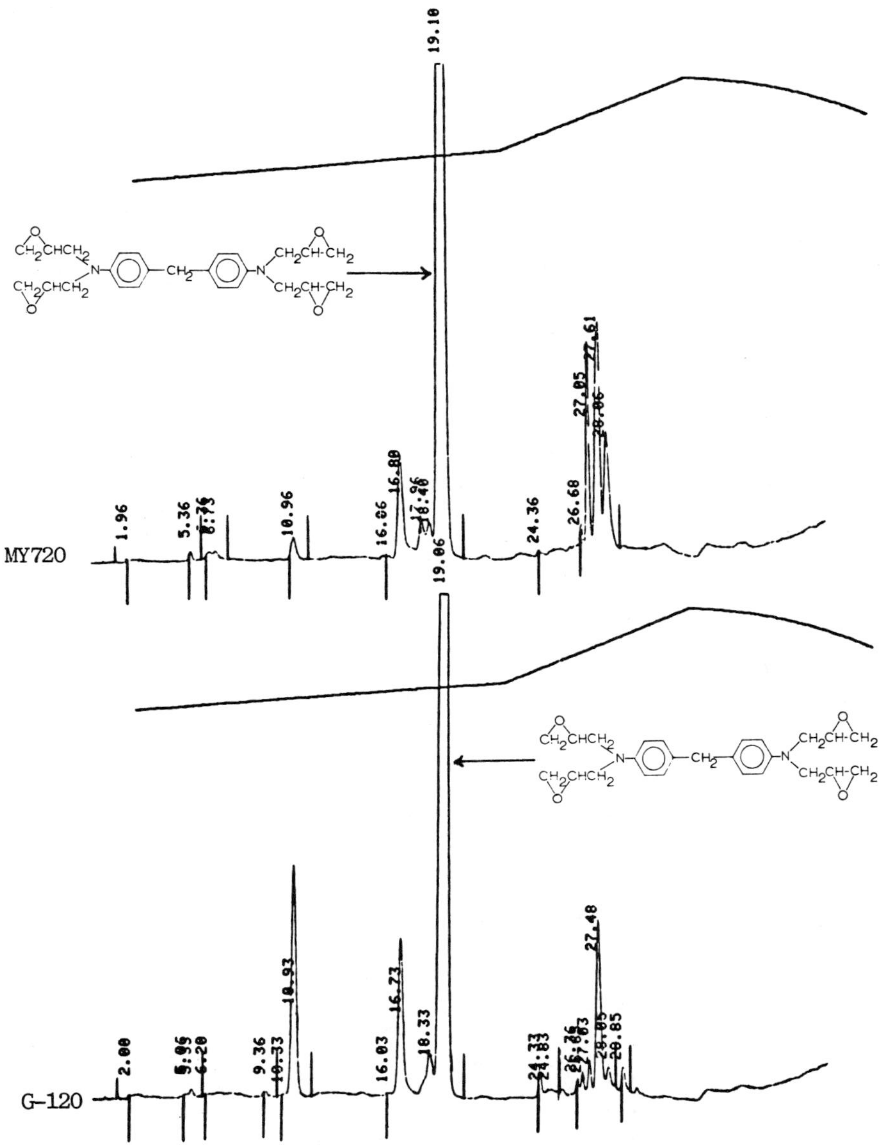

Figure 2. Reverse Phase HPLC of Commercial TGMDA Resins.

oligomeric components have retention times higher than TGMDA; whereas more polar components with hydrolyzable chlorides and free hydroxyls tend to have shorter retention times. It is noted that although both samples appear to have the same types of components, the relative amounts of components in each sample are quite different. The effects of such components on the properties of TGMDA will be discussed in this section.

The EEW and % epoxy values determined by titration, % TGMDA calculated from the HPLC analyses, and viscosities of several TGMDA samples are compared in Table I. Samples H, G, D and E were obtained by preparative liquid chromatography. Sample D resulted from Stage 1 of a marginally successful separation where the column was overloaded. Samples G and H were isolated from Stages 1 and 2, respectively, of the purification procedure. Sample E consisted primarily of higher oligomers and was the material which remained on the column in Stage 1 and was stripped from the column with THF.

The % epoxy values are based upon the theoretical EEW value 105.5 g/eq for the TGMDA monomer. The resin G-120 has a larger TGMDA concentration and a higher % epoxy than MY720. The viscosity of the samples increases with EEW and depends on the concentration of the higher molecular weight components. Sample H is water clear, G is pale yellow, D is light amber, and the other samples are dark amber in color.

Table I

SAMPLE	EEW (g/eq)	% EPOXY	% TGMDA (HPLC) SIZE EXCLUSION	REVERSE PHASE	VISCOSITY (CP) 50°C
H	107	99	99.0	98.9	-
G	108	98	98.5	97.7	1270-1590
D	111	95	87.7	88.0	-
G-120	118	89	76.2	77.7	5000-1500*
MY720	127	83	61.5	69.2	17600
E	168	63	-	9.0	50400-54700

*Manufacturer's specifications

Size exclusion and reverse phase HPLC chromatograms for some of the samples are compared in Figures 3 and 4. It is noted that samples H and G have a single major component; i.e., the TGMDA monomer. Preparative LC concentrates the minor components and leaves only about 9.4% of the TGMDA monomer in sample E.

Appreciable differences are also observed upon comparing the FTIR spectra of samples H, G-120 and E (Figure 5). The relatively sharp IR absorption bands, the large epoxy absorbance at 905cm^{-1} and the absence of an hydroxyl band are noted in the spectrum of sample H. Sample E has a large hydroxyl absorption band and an epoxy absorbance nearly half the size of that observed for sample H. In comparison with sample H, the

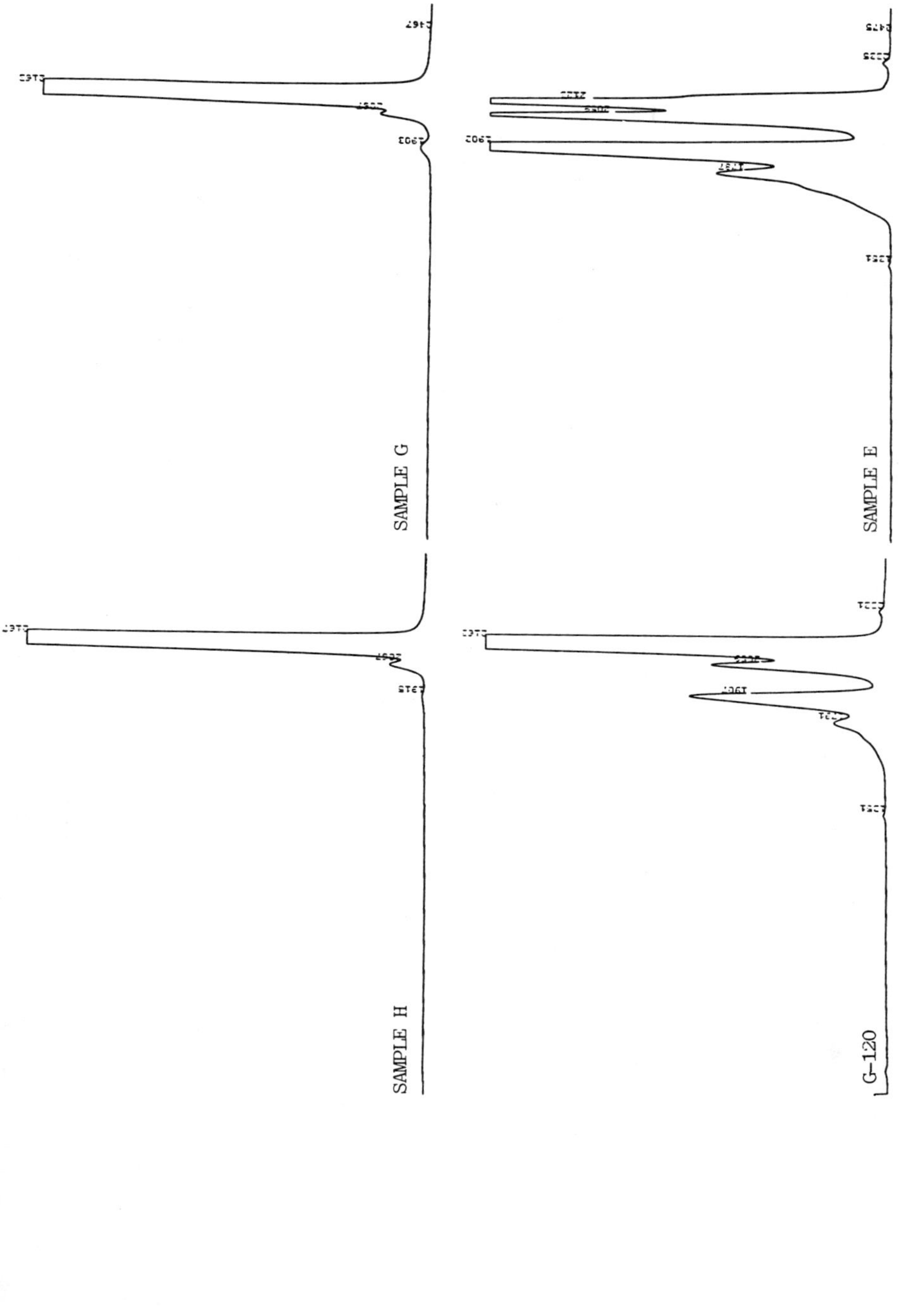

Figure 3. Size Exclusion HPLC of TGMDA Samples.

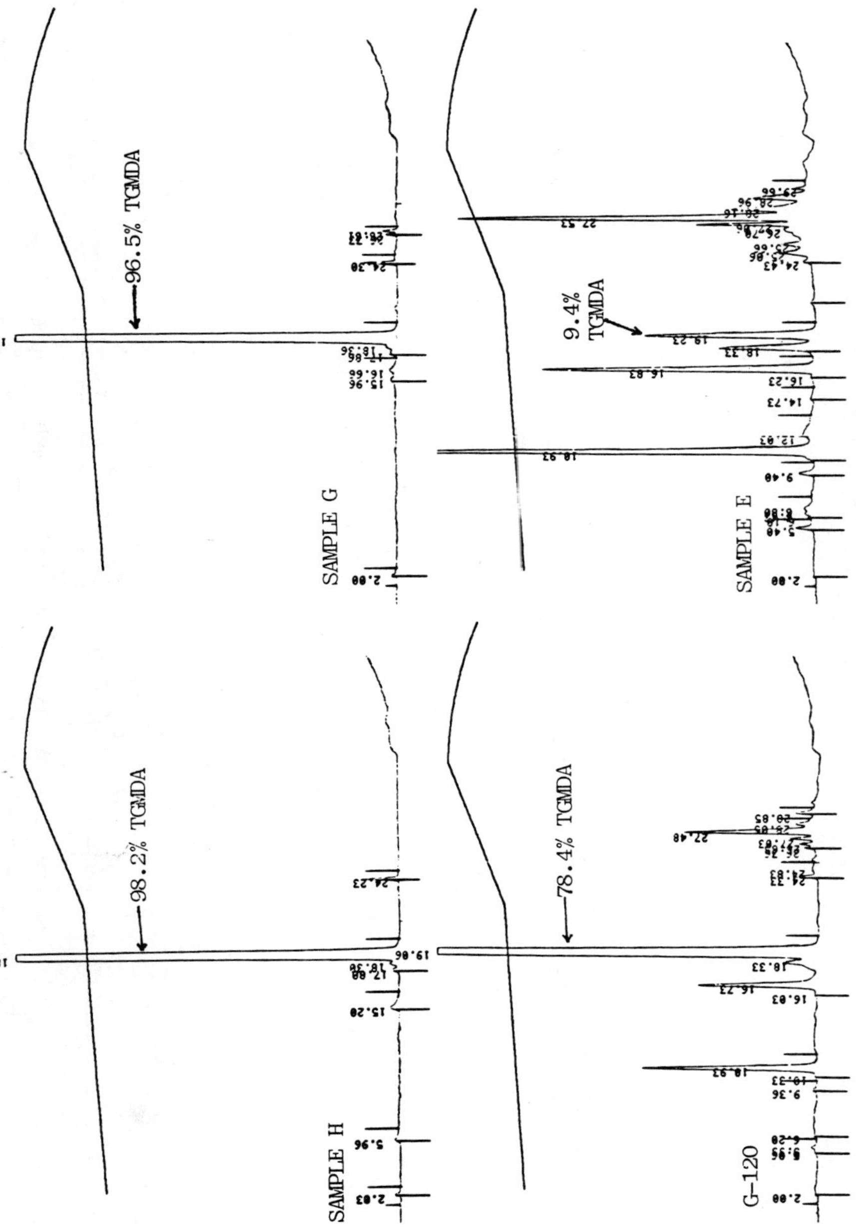

Figure 4. Reverse Phase HPLC of TGMDA Samples.

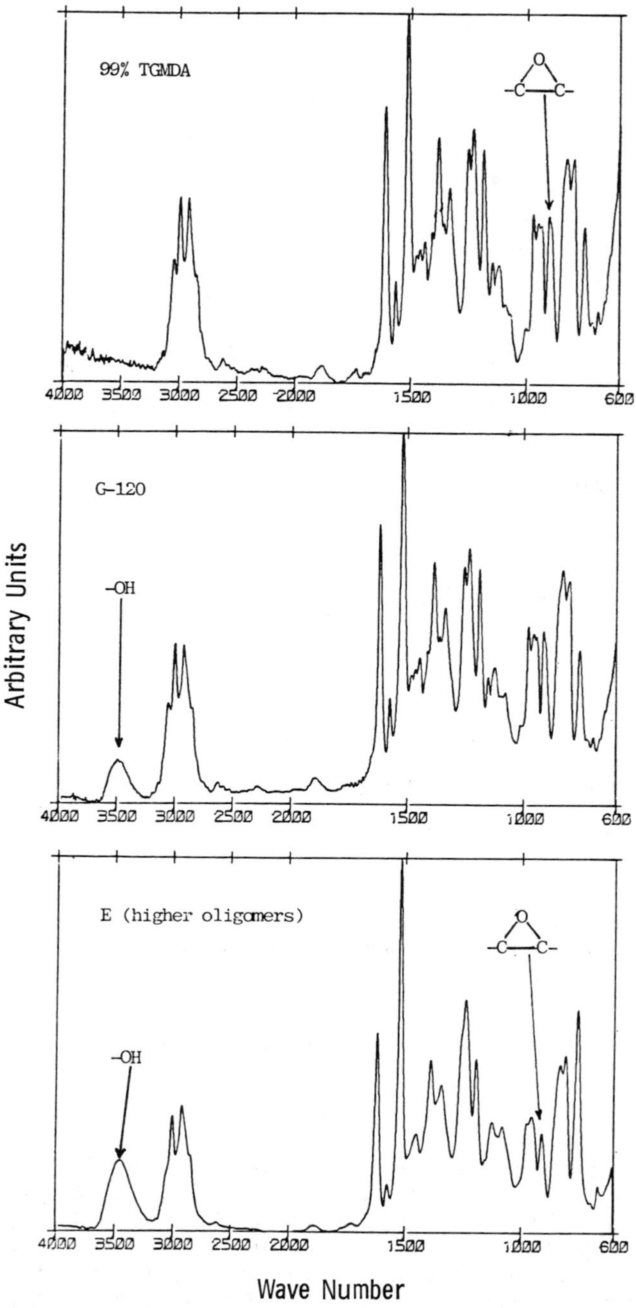

Figure 5. FTIR Spectra of TGMDA Samples H, G-120 and E.

absorption bands for sample E are broader and poorer resolved which suggest that sample E is heterogeneous and more complex structurally.

The purified TGMDA samples are found to be considerably more hydrolytically stable than the commercial resins. In the early stages of hydrolysis, the major product is identified as partially hydrolyzed TGMDA with three epoxy groups unreacted and one epoxy group ring-opened to form an α-glycol. This product is noted as a peak (11.1 minutes) in the HPLC chromatograms (Figure 6). As the samples age at 60°C and 96% RH, the product forms and the peak grows until the resins start crosslinking and become insoluble. The weight percentage of this product is plotted as a function of exposure time in Figure 7. The hydrolytic stabilities of the fresh MY720 and G-120 samples are quite similar; however, an older sample of MY720 (obtained in 1978) was found to be significantly less stable than the recent batch eventhough they initially had nearly identical % TGMDA. Over an aging period of one week, the % TGMDA in sample G changes very slightly compared to the changes in the commercial resins (Table II).

Table II
Hydrolytic Stability of TGMDA Resins

60°C, 96% RH

Sample	Time to Form 2% Hydrolysis Product	% TGMDA 0	% TGMDA 7 days	k_1 (day^{-1})
G (96% TGMDA)	15 days	98	96	0.084
G-120	1	76	59	0.35
MY720 (new)	2.1	66	47	0.37
MY720 (old)	0.4	66	44	0.35

The rate of formation of the hydrolysis product has a first-order dependence on the concentration of the hydrolysis product (Figure 8).

$$\frac{d\,[\text{hyd prod}]}{dt} = k_1\,[\text{hyd prod}]$$

$$\ln\,[\text{hyd prod}] = k_1 t$$

During the first day of exposure there is a rapid increase in the concentration of the hydrolysis product which does not follow the first-order relationship and which may be due to the hydrolysis of monomer with hydrolyzable chlorides. After the first day, the first-order rate expression applies until the concentration of the hydrolysis product becomes significant to the extent that it undergoes further hydrolysis and until an insoluble gel starts

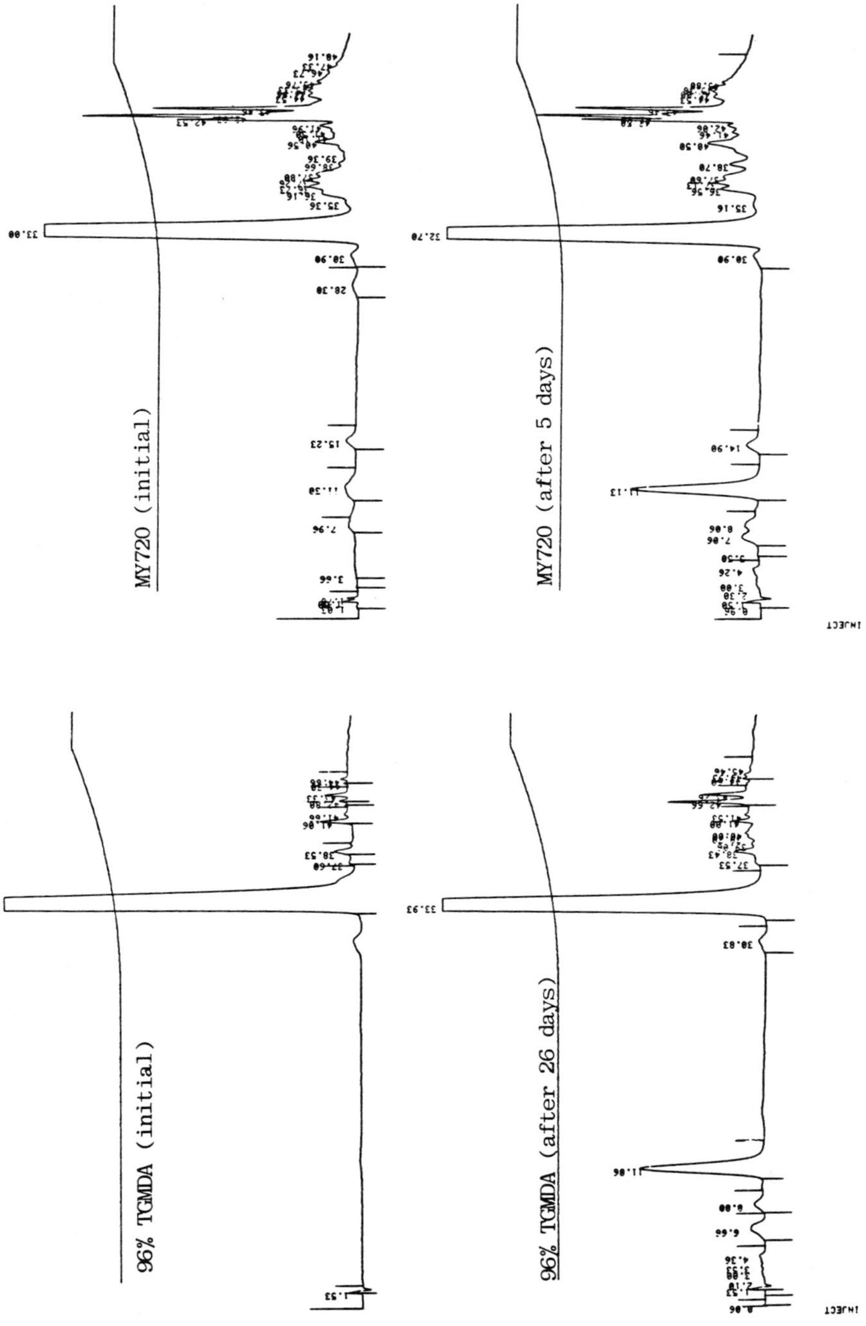

Figure 6. Hydrolytic Stability of TGMDA Resins at 60°C and 96% RH. Reverse Phase HPLC Analysis.

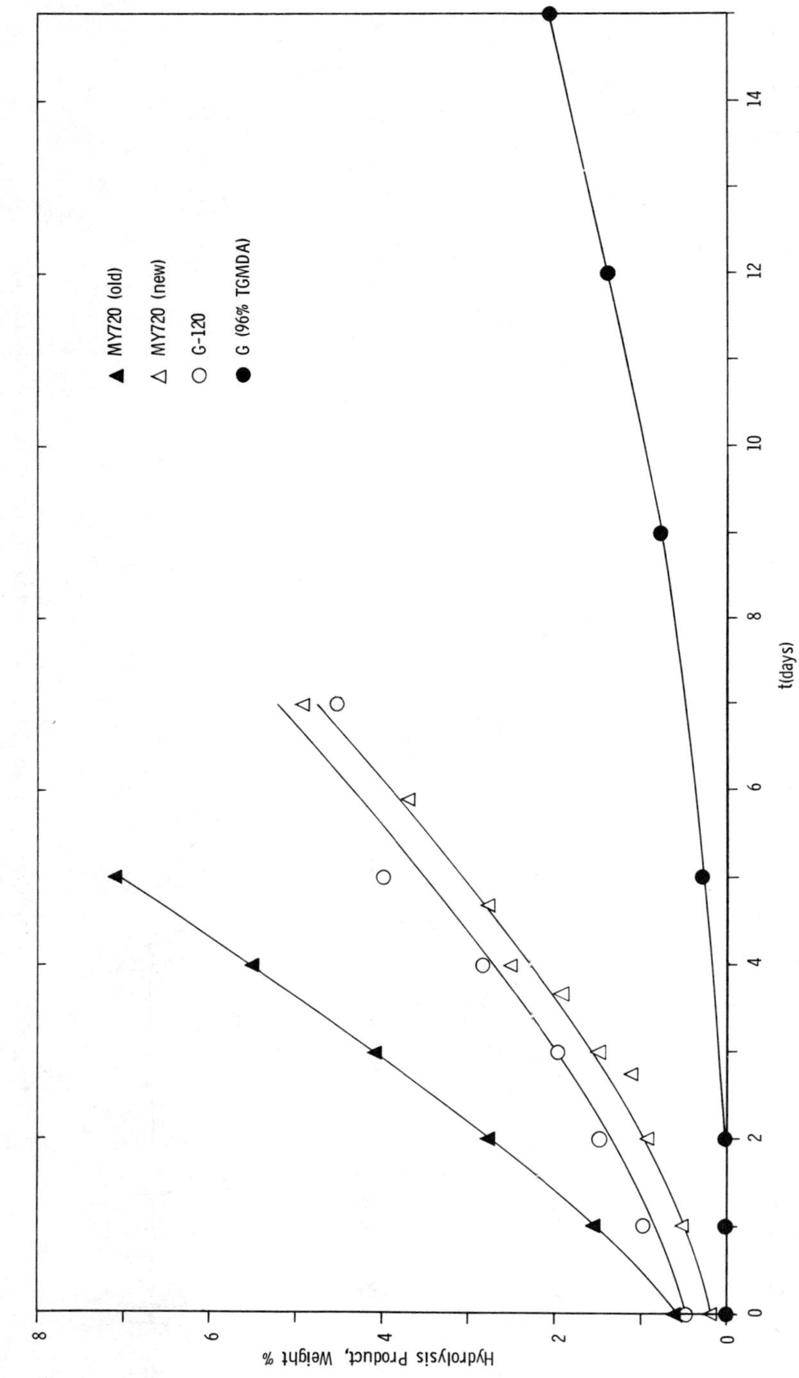

Figure 7. Hydrolytic Stability of TGMDA Resins at 60°C and 96% RH.

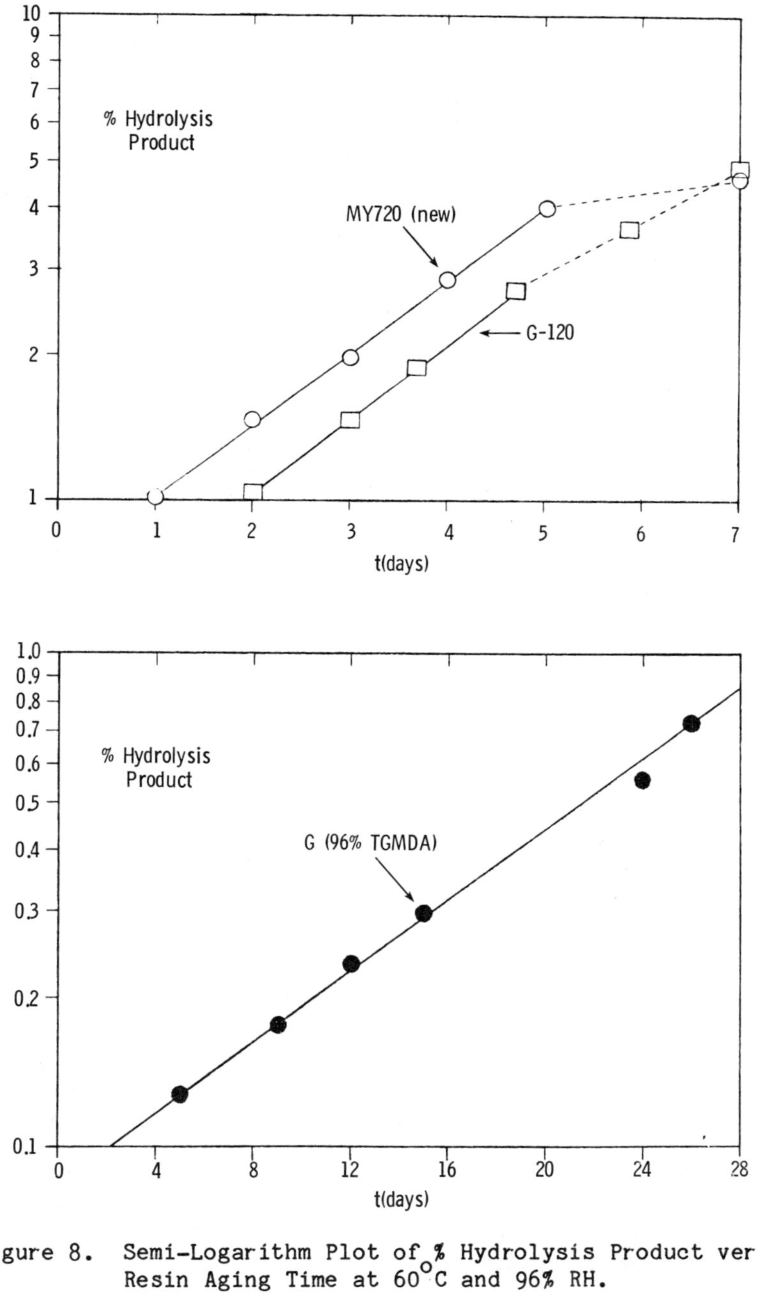

Figure 8. Semi-Logarithm Plot of % Hydrolysis Product versus Resin Aging Time at 60°C and 96% RH.

forming. The rate constants of the commercial resins are nearly identical (Table II). The smaller constant for the purified resin suggests that the hydrolysis mechanism is complex and is affected by the presence of minor components, in addition to the hydrolysis product analyzed, behaving as catalysts.

Effects of resin purity are observed in the thermal polymerization of the TGMDA samples and in the curing behavior of TGMDA/DDS mixtures. Characteristic exotherm curves are generated using DSC (5°C/min). Typically, the thermal polymerization reactions occur over a narrow temperature range and have sharp exotherm peaks. In comparison, the shape of the DSC exotherm curve and temperature range of the TGMDA/DDS (20 wt%) curing reaction is quite sensitive to the presence of impurities (Figure 9). For both reactions the exotherm temperature maximum T_{exo} and the enthalpy $-\Delta H$ decrease as the concentration of impurities or of components other than TGMDA increase (Table III). The ΔH value is directly related to the % epoxy and the decrease in T_{exo} is evidently caused by impurities behaving as catalysts.

Table III
DSC Analysis (5°C/min, N_2)

Sample	Thermal Polymerization of TGMDA Samples			Curing Reaction of TGMDA-DDS (20%) Mixtures		
	$-\Delta H$(cal/g)	T_{exo}(°C)	Tg(°C)	$\Delta H/\Delta H$(99%)	T_{exo}(°C)	Tg(°C)
H (99%)	314	309	134	1	245	244
G (97%)	316	310	145	0.92	238	240
D (88%)	312	304	156	0.89	233	231
G-120 (78%)	303	291	158	0.83	237	212
MY720 (69%)	279	296	168	0.84	232	202
E (9%)	225	265	181	0.62	228	181

Impurities have a significant effect on the glass transition temperatures Tg of the polymerization products (Table III). With increasing amounts of impurities, the Tg of the thermal polymerization products increase; whereas the Tg of the DDS curing reaction products decrease. It is noted that the Tg of the thermal polymerization product of sample H is 110°C lower than the value of the curing reaction product. The effect of impurities on the Tg of the DDS curing reaction product can be understood as a consequence of impurities decreasing the average functionality of reactive species while tending to catalyze epoxy homopolymerization. The effect of impurities on the Tg of the thermal polymerization products is more difficult to understand. Conceivably, homopolymerization involving the intramolecular reaction of diglycidyl groups may occur and thereby lower the crosslink density and hence the Tg of the thermal polymerization product to a greater extent as the TGMDA resin becomes more pure. Impurities, such as those containing α-glycol groups, may enhance

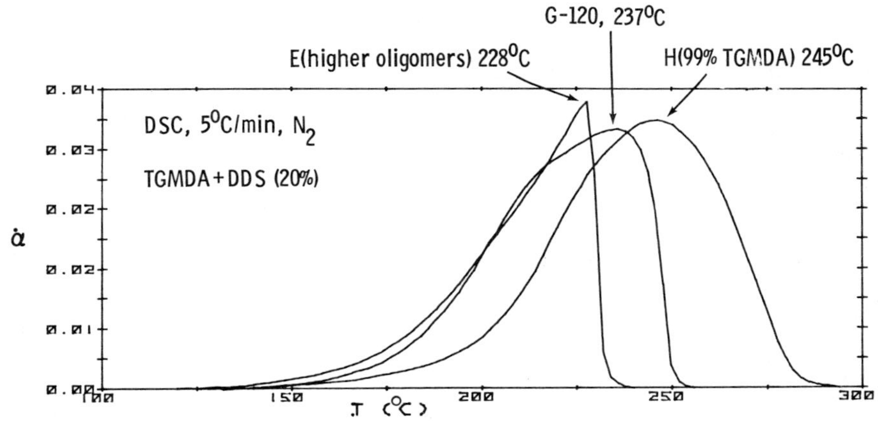

Figure 9. DSC Analysis of TGMDA/DDS Curing Reaction.

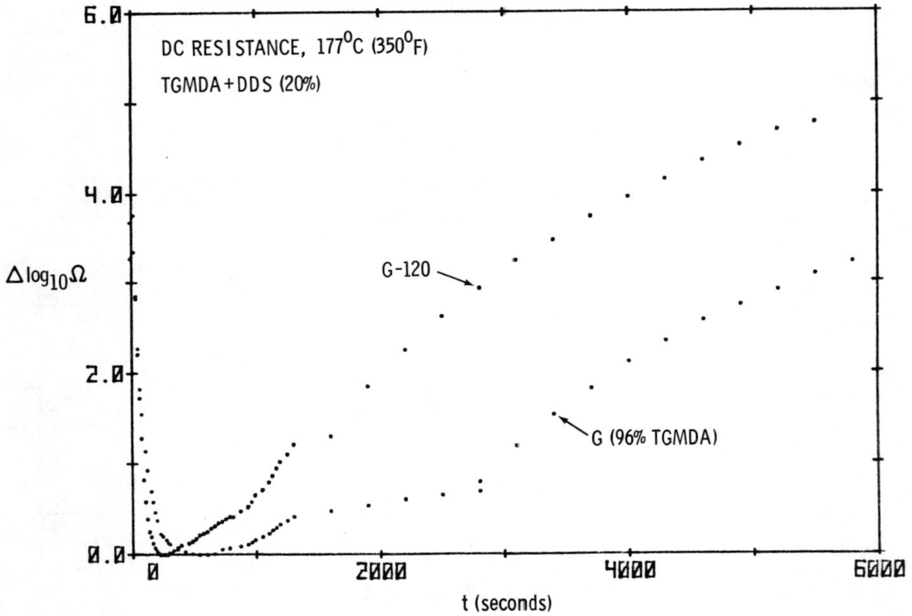

Figure 10. DC Resistance Measurements of TGMDA/DDS Curing Reaction.

the average functionality by initiating polymerization which would tend to raise the crosslink density and, therefore, the Tg of the products of the resins which are less pure. Since all the thermal polymerization samples were prepared using the same heating cycle, it is unlikely that degradation would have occurred to the extent to account for the lower Tg of the product obtained from highly pure TGMDA. FTIR studies (2) of the thermal polymerization may help explain the effect of impurities on Tg.

DC resistance studies of the TGMDA/DDS curing reaction indicate that impurities affect processability. Changes in electrical resistance Ω(ohms) are related to the mobility of the charge carriers and hence to the viscosity of the reaction medium. As the reaction mixture approaches the reaction temperature, the DC resistance measured between two electrodes achieves a minimum value Ω_{min}. Onset of the curing reaction generates an exotherm which upsets thermal equilibrium; however, as curing proceeds the reaction mixture becomes increasingly more viscous, especially with the onset of gelation, and the resistance increases. In Figure 10 $\Delta\log_{10}\Omega = \log_{10}\Omega - \log_{10}\Omega_{min}$ is plotted versus curing time t for commercial and purified TGMDA/DDS (20 wt%) mixtures at 177°C. The commercial sample starts reacting earlier and shows a much greater increase in Ω with reaction time than the purer sample. The plot for sample G has a peculiar steplike characteristic which may be related to a difference in polymerization mechanism or perhaps to an experimental anomaly.

This study has shown that synthesis by-products or impurities in TGMDA resins have a significant effect on resin properties and curing behavior. Future work will involve identifying and evaluating the effects of specific impurities. Preparative liquid chromatography will be used to purify large quantities of TGMDA to determine the effects of impurities on rheology and on the mechanical properties of the cured resin.

Acknowledgment

The authors are grateful to Mr. Walter X. Zukas for the viscosity measurements, to Mr. Bernard R. LaLiberte and Mrs. Emily McHugh for running the DSC analysis and to Dr. Richard J. Shuford for providing the DC resistance measurements.

Literature Cited

1. Jay, R. R., Anal. Chem., 1964, 36, 667.
2. Mones, E. T.; Morgan, R. J., Polym. Prep., Am. Chem. Soc., Div. Polym. Chem., 1981, 22(2), 249-250..

RECEIVED December 16, 1982

11

Structure–Property Relations of Polyethertriamine-Cured Bisphenol-A-diglycidyl Ether Epoxies

FUNG-MING KONG, CONNIE M. WALKUP, and ROGER J. MORGAN

Lawrence Livermore National Laboratory, Livermore, CA 94550

The relations between the chemical and physical network structure, the deformation and failure processes and the tensile mechanical properties of polyethertriamine cured bisphenol-A-diglycidyl ether epoxies are reported for a series of epoxy glasses prepared from a range of polyethertriamine concentrations. Near infrared spectroscopy indicates that these glasses form exclusively from epoxide-amine addition reactions. Their T_g exhibits a maximum and their swell ratio a minimum at the highest crosslink density. Stress-birefringence studies reveal that these highly crosslinked glasses are ductile and undergo necking and plastic deformation. The plastic deformation initially occurs homogeneously, but ultimately becomes inhomogeneous and shear bands develop. Tensile failure occurs in the high strain shear band region. The ultimate tensile strain of these epoxies attains a maximum of 15% for the highest crosslinked glass. Off-stoichiometric networks fail at lower strains because such networks inherently contain more defects in the form of unreacted ends. The density, yield stress, tensile strength, and modulus of these glasses all decrease with increasing polyethertriamine concentration as a result of increasing free volume because of the poor packing ability of the amine molecule. A minimum is superimposed on this downtrend in density and mechanical properties with increasing amine content at the highest crosslink density because of geometric constraints imposed on segmental packing by the network crosslinks. The ability of these crosslinked glasses to undergo deformation is discussed in terms of the free volume and the crosslinked network topography. Network failure is considered in terms of stress-induced chain scission which is determined by the concentration and extensibility of the least extensible network segments.

0097–6156/83/0221–0211$06.00/0

The increasing use of high performance, fibrous composites in critical structural applications has led to a need to predict the lifetimes of these materials in service environments. To predict the durability of a composite in service environment requires a basic understanding of (1) the microscopic deformation and failure processes of the composite; (2) the roles played by the fiber, epoxy matrix and fiber-matrix interfacial region in composite performance; and (3) the relations between the structure, deformation and failure processes and mechanical response of the fiber, epoxy matrix and their interface and how such relations are modified by environmental factors.

In this paper we address the area of the structure-property relations of epoxy matrices, with specific emphasis on amine-cured epoxides. In attempts to correlate the structure-property relations of amine-cured epoxides, there have been a number of studies on the mechanical properties of these glasses as a function of epoxide:amine ratios and the chemical structure of the constituent epoxide and amine monomers.(1-15) Generally, there is no direct correlation between the chemistry of the epoxide and amine monomers and the mechanical properties of epoxies with the exception that as the distance between crosslinks becomes shorter, these glasses become more brittle.(1,10,15) Also, for a specific amine-cured epoxide system, the T_g is always highest for the fully reacted, highest crosslink density glass.(3,6,9,12, 16,17,18)

There are a number of reasons why there is not a simple correlation between the chemistry of the starting materials and the mechanical properties of the resultant cured epoxy glass. First, the chemistry of the epoxy network may not always be simple. Amine-cured epoxy networks are generally assumed to result exclusively from addition reactions of epoxide groups with primary and secondary amines.(16) However, these reactions are often incomplete due to steric and diffusional restrictions, and additional reactions such as epoxide homopolymerization can occur.(8,12,13, 16,19-28) The physical structures of epoxies are also complex on a macroscopic and microscopic scale and have not been characterized in detail. Macroscopically, the crosslinked networks are inhomogeneous because of voids and possible distributions in the crosslink density. On a more microscopic level, details of the local free volume, the rotational isomeric configurations of segments between crosslinks, chain end defects, and any bond angle and length distortions produced to form the crosslinked networks are generally unknown. These chemical and physical structural parameters play an intimate role in the deformation and failure processes and mechanical response of the glassy crosslinked networks. However, few studies have been conducted on the microscopic deformation and failure processes of epoxy glasses.(13, 27,29,30) Therefore, at present there are considerable gaps in our understanding of the structure-property relations of epoxies.

To further our understanding of crosslinked epoxies, in this paper we report structure-property studies of a series of polyethertriamine cured bisphenol-A-diglycidyl ether (DGEBA) epoxies whose chemical structures are well characterized. Different epoxy networks were produced by varying the epoxide:amine ratios. The physical structure, deformation and failure processes and mechanical properties of these glasses are reported and are correlated with the systematic chemical structural changes in this epoxy series. The high ductility of these epoxies and their relative insensitivity to inherent fabrication flaws were advantageous in correlating their network structure with their mechanical response.

Experimental

Materials. Pure DGEBA, DER 332 (Dow) epoxide monomer was used in this study and was cured with an aliphatic polyethertriamine, Jeffamine T403 (Jefferson). The chemical structures of the amine and epoxide monomers are shown in Figure 1. The epoxide and amine equivalent weights determined by chemical titration were 175.15 and 82.92 grams, respectively. If these systems form exclusively from epoxide-amine addition reactions, about 47 phr (parts per hundred) T403 are required for complete cure. A range of epoxies with 25 to 75 phr T403 was prepared. The DGEBA epoxide monomer was preheated to 60°C to melt any crystals present(8) prior to mixing with the T403. After mixing the two monomers, the mixtures were vacuum degassed and cast between glass plates separated by 0.3 cm thick Teflon spacers. A thin layer of release agent (DC-20 Dow) was baked onto each plate prior to casting. The epoxy sheets were cured between the glass plates at room temperature for 24 hours, post-cured at 85°C for 16 hours and then slowly cooled at about 2°C/min to room temperature. All specimens used for physical structural characterization and mechanical property tests were fabricated from these epoxy sheets.

Experimental. Near infrared spectroscopy (Cary 14 spectrophotometer) was used to study the cure reactions and the resultant chemical structure of the epoxies. A correct portion of epoxide and amine were mixed, degassed, and poured into a glass cell. To reduce the path length, a thick glass plate was inserted. A copper constantan thermocouple was used to control the test temperature. Each epoxy mixture was scanned at room temperature for 24 hours and at 85°C for 16 hours. The absorbances of the epoxide (A_e) and phenyl (A_p) groups were monitored at 2.205 and 2.160 μm, respectively. Since the phenyl group does not participate in any of the chemical reactions, this group's absorbance was utilized as an internal standard. The epoxide group consumption was determined by monitoring the changes in the A_e:A_p ratio with cure time.

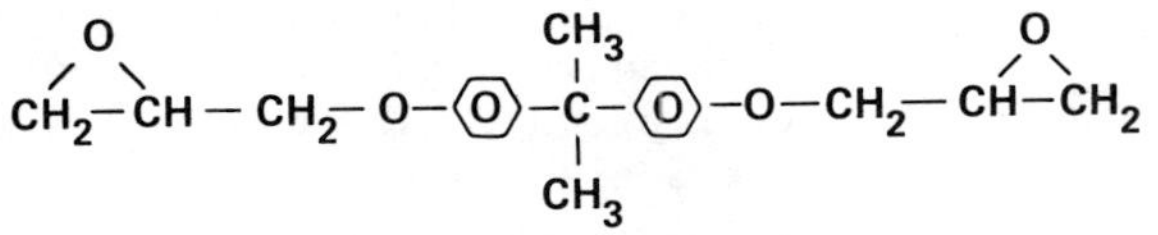

Diglycidyl ether of bisphenol A
DER 332

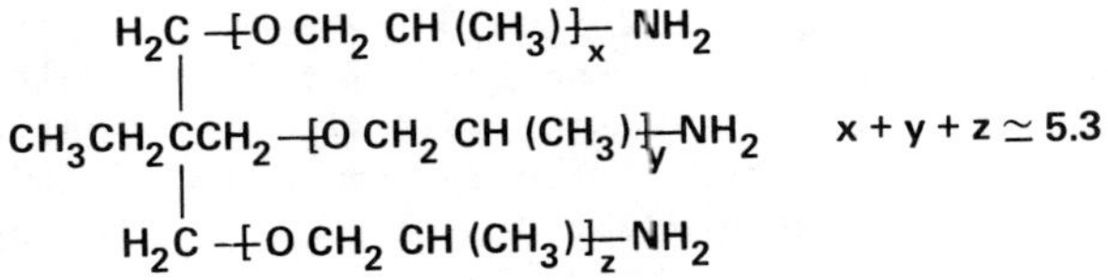

Polyether triamine
T403

Figure 1. Chemical structures of DGEBA epoxide and T403 polyethertriamine monomers.

A differential scanning calorimeter (DSC) (Du Pont 910) was used to monitor the T_g's of the epoxies. The heating rate was 10°C/min. The dynamic mechanical properties of the epoxies were monitored by a Rheometrics mechanical spectrometer (Model RMS-7200). Epoxy specimens with dimensions of 6.4x1.3x0.3 cm were torqued at an oscillating frequency of 2.0 Hz. The shear storage (G') and loss (G") moduli and tan δ were determined from -160°C to +140°C. The densities of the epoxies were performed on 1x1x0.3 cm epoxy specimens that were immersed in methyl ethyl ketone for seven days. The weights (w) and volumes (v) of the specimens were measured prior to (w_i and V_i) and after (w_f and v_f) immersion and the swelling ratio v_f/v_i determined.

For the mechanical tensile tests, dogbone-shaped specimens with a gauge length 7.5 cm and a width of 1.3 cm were cut from the cast epoxy sheets. After polishing the specimen edges along the gauge length, the specimens were annealed at 85°C under vacuum for 1.5 hours to remove any sorbed moisture and release any fabrication strains. The specimens were stored in a dessicator prior to testing. All tensile tests were conducted at room temperature on a mechanical tester (Instron, TTDM) at a crosshead speed of 0.5 cm/min. The fracture topographies of the failed specimens were documented by optical microscopy (Zeiss).

The deformation modes were monitored with a polariscope (Photoelastic Inc.) as a function of stress for epoxies with 37, 47, and 57 phr T403. An attached camera was used to record the birefringence patterns.

Results and Discussion

Chemical Structure. Amine-cured epoxy networks generally form exclusively from epoxide-amine addition reactions; i.e.,

$$R_1\underbrace{CH-CH_2}_{O} + R_2NH_2 \longrightarrow R_1CH(OH)CH_2NHR_2$$

For all the amine hydrogens to react with epoxide groups in the DGEBA-T403 system exclusively by epoxide-amine addition reactions would require a stoichiometric quantity of 47 phr T403 based on the chemically determined equivalent weights of the DGEBA and T403 reactants.

The epoxide consumption during cure was monitored by near IR as a function of T403 concentration (25-75 phr range) for a 24 hour cure at 23°C followed by a postcure at 85°C. The IR measurements were truncated when either a 100% epoxide consumption occurred or the epoxide consumption ceased to change with time at 85°C. The ultimate percentage epoxide consumptions at 85°C are plotted as a function of T403 concentration in Figure 2. All epoxide groups are consumed at $\geq$45 phr T403. The lowest T403 concentration of 45 phr for full epoxide consumption is slightly lower than the 47 phr T403 predicted from chemical analysis of

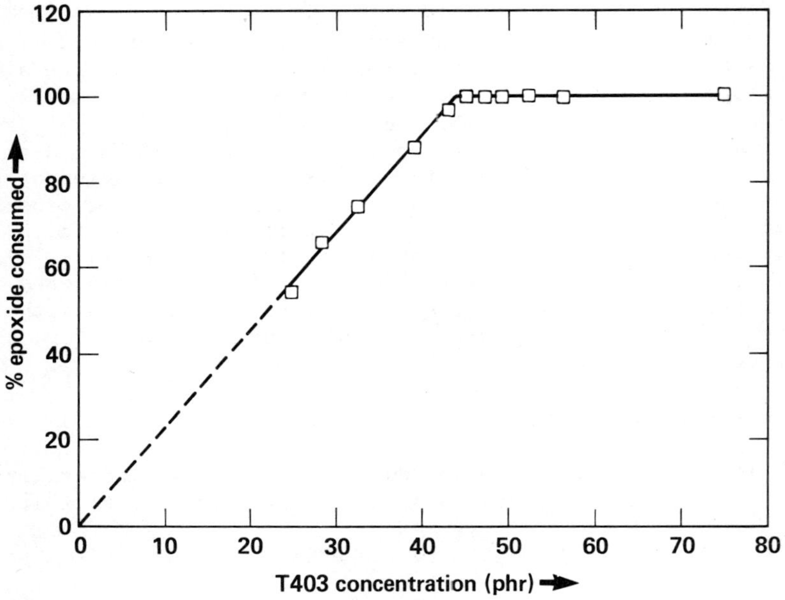

Figure 2. Ultimate percent epoxide consumption versus T403 concentration for 24-hour cure at 23°C followed by 16-hour 85°C postcure for DGEBA-T403 epoxies.

the starting materials and assuming exclusively epoxide-amine addition reactions. This slight discrepancy could be due to insufficient accuracies in (1) the chemical titration procedures to determine the DGEBA and T403 equivalent weights and/or (2) in the near IR to detect small differences in epoxide concentrations. In the 25-45 phr T403 range in Figure 2, the epoxide consumption levels fall on a straight line which, when extrapolated, go through the origin. This data indicates that the DGEBA-T403 epoxies form exclusively from epoxide-amine addition reactions which proceed to completion without any additional epoxide consumption occurring as a result of side reactions.

Network Structure. The physical network structures and the average molecular weight between crosslinks, M_c, for DGEBA-T403 epoxies as a function of T403 concentration were determined and are illustrated in Table I. These structures were ascertained based on the chemical and physical structure of the DGEBA and T403 monomers and the following assumptions: (1) the cure reactions and network formation result exclusively from epoxide-amine addition reactions; (2) all primary amine hydrogens react with epoxide groups prior to the onset of secondary amine-epoxide addition reactions.

The functionalities and flexibilities of the DGEBA and T403 structures allow a variety of ring structures to form in the fully crosslinked DGEBA-T403 (~47 phr T403) epoxy resulting in an irregular network. Molecular models indicate such rings are relatively deformable. However, because of the irregular network, cooperative ring deformations will be severely restricted and indeed lead to network chain scissions. The network topography is even more irregular than illustrated in Table I because of the different lengths of the three arms of the T403 molecule in which 23% have an M_c value of 200 and the remainder an M_c value of 158. These M_c values are determined by assuming x, y, or z = 1 or 2 in the T403 chemical structure illustrated in Figure 1. Molecular models also indicate that the packing efficiency of the network segments is the most severely restricted for the highest crosslink density network.

A plot of M_c vs T403 concentration, determined from the cure reactions and epoxide and amine equivalent weights, is illustrated in Figure 3 as the full line. The experimental data points in this figure were from Nielsen's (31) empirical equation

$$\log_{10} G' \simeq 7.0 + 293d/M_c$$

where G' is the shear modulus above T_g in dynes/cm^2 and d is the polymer density. The G' values were determined from dynamic mechanical measurements above T_g in the 100-120°C range. The experimental data points follow the same general trend as the chemically determined M_c-T403 concentration plot but exhibit on the average ~30% larger M_c values. The agreement between

Table I. The Network Structures and Molecular weight, Mc, for DGEBA-T403 Epoxies

T403 Concentration phr	Structure	Comments	M_c
23.7	(a) (b) 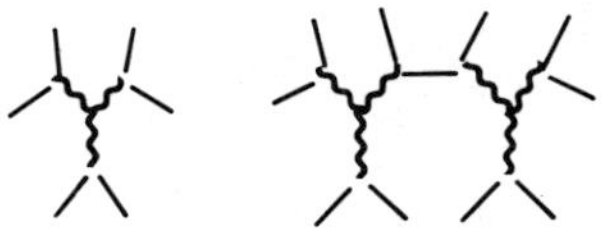	All-NH_2 and -NH groups react; (a) one $-C\overset{O}{-}C-$ group reacts on each DGEBA molecule; (b) In some DGEBA molecules both $-C\overset{O}{-}C-$ groups reach whereas in others no $-C\overset{O}{-}C-$ groups react	∞
47.3		All -NH_2, -NH and $-C\overset{O}{-}C-$groups reacted completely A network of interconnected rings	258
94.7	(a) (b) 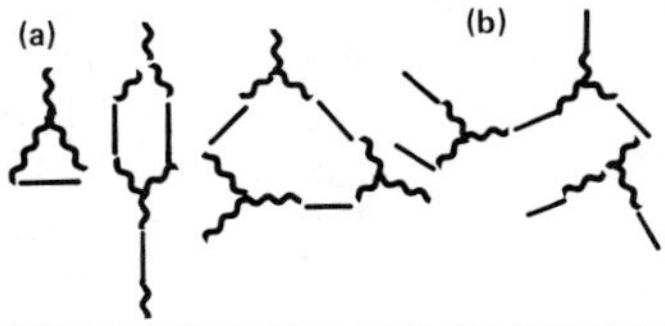	All -NH_2 and $-C\overset{O}{-}C-$ groups react; no -NH groups react. (a) Ring and (b) branched structures	682
142.0	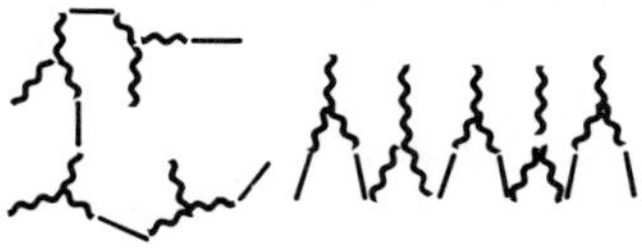	All $-C\overset{O}{-}C-$ groups react; 2/3 NH_2 groups react; no NH groups react; (a) branched and (b) linear structures	∞
	T403  DGEBA —		

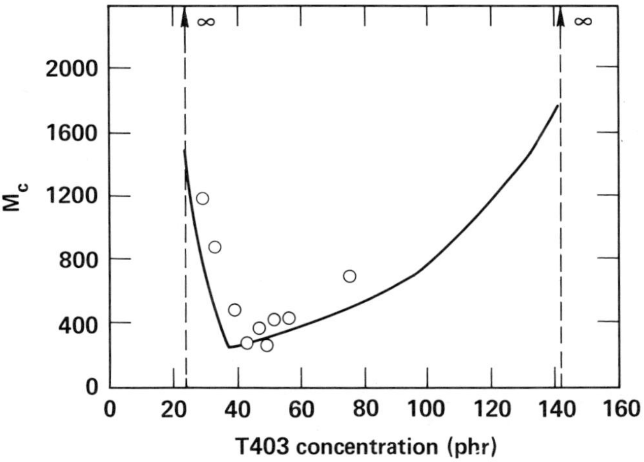

Figure 3. M_c versus T403 concentration for DGEBA-T403 epoxies, determined from cure reactions (Experimental data points determined from G' values above T_g).

the data is relatively good in the light of the empirical nature of the M_c-$\log_{10}$ G' relationship. A more rapid increase in M_c with T403 concentration occurs on the excess epoxide of stoichiometry relative to the excess T403 side.

Physical Properties. The physical properties of the DGEBA-T403 epoxies as a function of T403 concentration can be understood in terms of the network structures discussed in the previous section.

In Figure 4, the T_g's of the DGEBA-T403 epoxies, determined from differential scanning calorimetry measurements, are plotted versus T403 concentration. The highest T_g, 93°C, is exhibited by the epoxy with the highest crosslink density at ~45 phr T403. The T_g decreases less sharply with T403 concentration on the excess T403 side of stoichiometry because M_c increases less sharply on this side of stoichiometry. The swell ratio as expected exhibits a minimum for the highest crosslinked glass.

A plot of the 23°C density of DGEBA-T403 epoxies vs T403 concentration, in Figure 5, reveals the density progressively decreases with increasing T403 concentration with a minimum superimposed on this trend near the T403 stoichiometric concentration. The inability of the three-armed T403 molecule to pack as well as the more linear DGEBA molecule causes a progressive density decrease with increasing T403 concentration. A minimum in the density is superimposed on this downtrend at the highest crosslink density because of the geometric constraints imposed on segmental packing by the crosslinked network geometry.

Mechanical Properties and Deformation and Failure Processes. The DGEBA-T403 epoxies are ductile in tension at 23°C and exhibit a macroscopic yield point, necking and cold drawing.

The ultimate strain of the DGEBA-T403 epoxies exhibits a maximum of 15% at the stoichiometric T403 concentration as shown in Figure 6. Below ~35 phr T403 the ultimate strain increases rapidly with decreasing T403 content as a result of increasing M_c, decreasing T_g and plasticization of the glasses by unreacted DGEBA molecules.

Above 35 phr T403 concentration, Young's modulus and the tensile macroscopic yield stress and strength all decrease with increasing T403 concentration. This downtrend is illustrated in Figure 7 for the macroscopic yield stress. A minimum is superimposed on this downtrend at the stoichiometric T403 concentration, which is most evident in the modulus data. Below 35 phr T403 concentration, these mechanical properties all rapidly decrease with decreasing T403 concentration as a result of increasing softening of the epoxy glasses.

The deformation processes of DGEBA-T403 epoxies were monitored under polarized light as a function of strain level. The birefringent patterns of DGEBA-T403 (47 phr T403) epoxy as a function of strain is illustrated in Figure 8. At strains

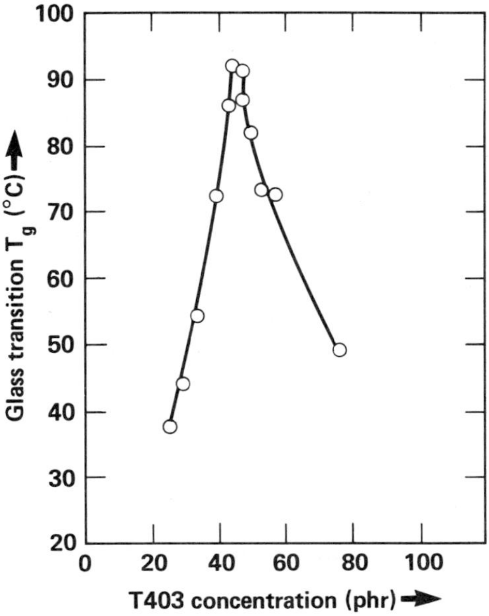

Figure 4. T_g versus T403 concentration for DGEBA-T403 epoxies.

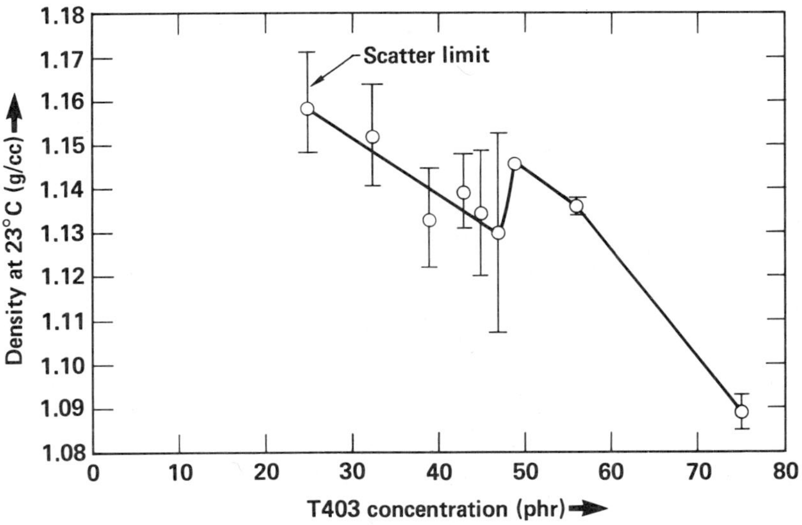

Figure 5. Density versus T403 concentration for DGEBA-T403 epoxies, at 23°C.

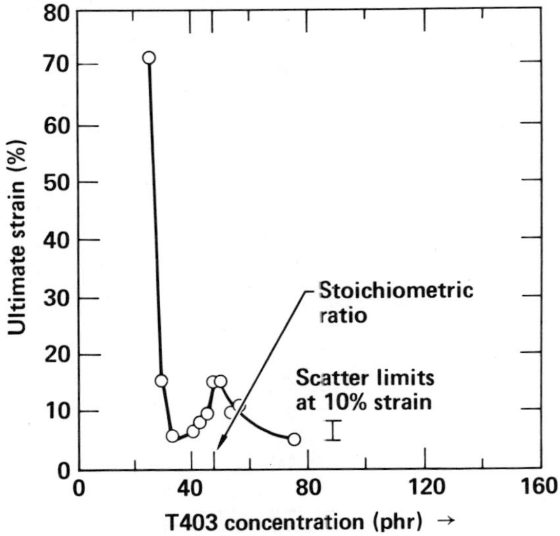

Figure 6. The ultimate tensile strain versus T403 concentration for DGEBA-T403 epoxies, at 23°C.

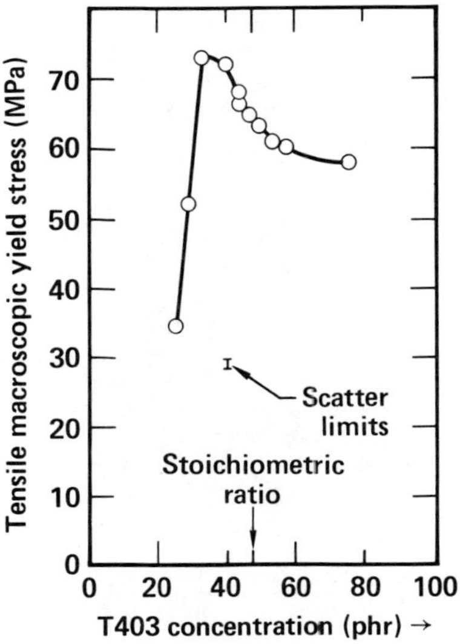

Figure 7. Tensile macroscopic yield stress of DGEBA-T403 epoxies as a function of T403 concentration, at 23°C.

Figure 8. Birefringent deformation processes in DGEBA-T403 (47 phr T403) epoxy as a function of strain at 23°C, under polarized light.

⋦2.5%, only elastic deformations occur and homogeneous color changes produced under polarized light in this strain region disappear instantaneously upon removal of the load. In the 2.5 to 4% strain region, homogeneous elastic and plastic deformations occur; and upon removal of the load the homogeneous plastic deformation does not relax out, which results in a permanent homogeneous color change in the unstrained epoxy when viewed under polarized light. Above 4% strain the plastic deformation becomes increasingly inhomogeneous with increasing strain as indicated by the development of birefringent fringes. Ultimately, a local region of high strain develops in the sample in the form of a diffuse shear band and a neck develops in this region. The birefringent fringes orient at ~45° to the direction of the applied load in the shear band. Fracture occurs in the high strain, shear band region. Fracture topography studies indicate failure occurs by a crude craze process, as has been observed in other epoxies.(30)

For the off-stoichiometric DGEBA-T403 epoxies that are in the 30-75 phr T403 range, birefringent-strain studies indicate the following trends in the deformation processes for epoxies that are increasingly further removed from the stoichiometric T403 concentration: (1) plastic flow is initiated and becomes inhomogeneous at lower strains; (2) the regions of high strain are less oriented and their associated bands are less well-developed upon crack propagation through such regions. DGEBA-T403 epoxies in the 20-30 phr T403 range that exhibit ultimate elongations in the 20-70% range deform only in a homogeneous plastic fashion.

Structural Factors Controlling Mechanical Properties and Deformation and Failure Processes. The ultimate elongation of the DGEBA-T403 epoxies is controlled by the concentration of inherent and/or stress-induced defects in the most highly strained region of the epoxy network. A maximum in the failure strain occurs at stoichiometry because the most highly cross-linked network contains the least number of defects. In fully reacted networks, network failure is initiated by chain scission in overstrained segments. Such scissions result in the neighboring segments carrying more of the applied load. Furthermore, the chemically active free radicals that form upon scission can react with neighboring segments to produce further scissions. Both these phenomena result in autocatalytic network degradation. The scission process will depend directly on the network extensibility and, therefore, on the network topography. The extensibility of the network segments in the DGEBA-T403 epoxy network will exhibit three discrete values as a result of there being three discrete M_c values. The DGEBA segments which make up 50% of the total network segments exhibit the largest M_c value of 342. The T403 segments have M_c values of 200 and 258 and are associated with the arms of the T403 molecule in which x, y, or z = 1 or 2, respectively, in Figure 2. Those T403 segments with the lowest M_c value of 200 are 11.5% of the total network segments. These least extensible segments carry a significant portion of the load, under stress, and will therefore be the first segment to undergo chain scission.

In addition to the segmental extensibility, the network topography will also play a significant role in the ultimate elastomeric network properties. The ability of the basic ring structures of the epoxy network to undergo deformation controls the network extensibility. The deformation of these rings depends on (1) the extensibility of their sides; (2) the flexibility of their internal angles; and (3) their ability to undergo cooperative deformation with their interconnected neighbors. Their cooperative deformation is controlled by the regularity of the network topography which is determined by the different geometries of the basic rings and their orientation relative to one another. The more regular networks consist of interconnected rings of similar size and shape. However, the more irregular networks consist of rings with a variety of geometries that will develop overstrained segments at lower extensibilities. The fully crosslinked DGEBA-T403 epoxy network is relatively irregular as illustrated in Table I.

The magnitudes of the macroscopic and microscopic yield stresses, the tensile strength (which is controlled by the flow properties of the epoxy) and Young's modulus E of DGEBA-T403 epoxies are determined by the glassy-state packing and free volume. These mechanical properties will follow the same trends as the density as a function of T403 concentration. The density of the DGEBA-T403 epoxies decreases with increasing T403 content with a minimum superimposed on such a downtrend at stoichiometry (Figure 5). Above ~35 phr T403 the yield stresses, tensile strength, and Young's modulus all follow the same trend as the density, and decrease with increasing T403 concentration. The minimum in the density at stoichiometry that is superimposed on the downtrend with increasing T403 concentration is revealed in the modulus data but is not clearly evident in the yield stress and tensile strength data. Below ~35 phr T403, the DGEBA-T403 epoxies become increasingly soft with decreasing T403 concentrations which causes sharp decreases in the yield stresses, tensile strength and modulus with decreasing T403 content, resulting in a maximum in such properties near 35 phr T403.

Conclusions

DGEBA-T403 epoxy glasses form exclusively from epoxide-amine addition reactions. Their T_g exhibits a maximum and their swell ratio a minimum at the highest crosslink density. These highly crosslinked glasses are ductile and undergo necking and plastic deformation. The plastic deformation initially occurs homogeneously, but ultimately develops inhomogeneously in the form of shear bands. Failure occurs in the high strain, shear band region. The ultimate strain of these crosslinked glassy networks is determined by the concentration of unreacted and/or broken network segments. Stress-induced chain scission of network segments is determined by the concentration and extensibility of the least extensible segments. The ability of these epoxies to undergo plastic flow depends on (1) the deformability of the basic network ring structures and their ability to undergo cooperative motion; and (2) the glassy-state free volume. The free volume depends on the molecular shape and packing ability of the constituent epoxide and amine portion of the network and the geometric constraints imposed on segmental packing by the network geometry.

Acknowledgments

This work was performed under the auspices of the U. S. Department of Energy by Lawrence Livermore National Laboratory under Contract No. W7405-Eng-48.

Literature Cited

1. Lee, H. and Neville, K., "Handbook of Epoxy Resins," McGraw-Hill, New York, 1967, Ch. 6.
2. Busso, C. J., Newey, H. A., and Holler, H. V., Proc. 25th Conf. of Reinforced Plastics/Composites Division of SPI, Washington, D.C., 1970.
3. Chiao, T. T., Jessop, E. S., and Newey, H. A., "A Moderate Temperature Curable Epoxy for Advanced Composites," Lawrence Livermore Laboratory Report UCRL-76126, 1974.
4. Chiao, T. T. and Moore, R. L., Proc. 29th Conf. of Reinforced Plastics/Composites Division of SPI, Washington, D.C., Section 16-B, 1, 1974.
5. Chiao, T. T., Jessop, E. S., and Newey, H. A., SAMPE Quarterly 1974, 6, No. 1.
6. Selby, K. and Miller, L. E., J. Mater. Sci. 1975, 10, 12.
7. Pritchard, G. and Rhoades, G. V., Mater. Sci. and Eng. 1976, 26, 1.
8. Morgan, R. J. and O'Neal, J. E., J. Macromol. Sci. Phys. 1978, 15, 139.
9. Kim, S. L., Skibo, M. D., Manson, J. A., Hertzberg, R. W., and Janiszewski, J., Polym. Eng. Sci. 1978, 18, 1093.
10. Phillips, D. C., Scott, J. M., and Jones, M., J. Mater. Sci. 1978, 13, 311.
11. Young, R. J. in "Developments in Polymer Fracture-1," Ed. E. H. Andrews, Appl. Sci. Publishers, London, 1979.
12. Morgan, R. J., J. Appl. Polym. Sci. 1979, 23, 2711.
13. Morgan, R. J., O'Neal, J. E., and Miller, D. B., J. Mater. Sci. 1979, 14, 109.
14. Yamini, S. and Young, R. J., J. Mater. Sci. 1980, 15, 1814.
15. Scott, J. M., Wells, G. M., and Phillips, D.C., J. Mater. Sci. 1980, 15, 1436.

16. Lee, H. and Neville, K., "Handbook of Epoxy Resins," McGraw-Hill, New York, 1967, Ch. 7.
17. Kreahling, R. P. and Kline, D. E., J. Appl. Polym. Sci. 1969, 13, 2411.
18. Murayama, T. and Bell, J. P., J. Polym. Sci. 1970, A-2, 8, 437.
19. French, D. M., Strecker, R. A. H., and Tompa, A. S., J. Appl. Polym. Sci. 1970, 14, 599.
20. Acitelli, M. A., Prime, P. B., and Sacher, E., Polymer 1971, 12, 335.
21. Prime, R. B. and Sacher, E., Polymer 1972, 13, 455.
22. Whiting, D. A. and Kline, D. E., J. Appl. Polym. Sci. 1974, 18, 1043.
23. Sidyakin, P.V., Vysokomol. Soyed 1972, A14, 979.
24. Tanaka, Y. and Mika, T. F., "Epoxy Resins," C. A. May and Y. Tanaka, Eds., Marcel Dekker, Inc., New York, 1973, Ch. 3.
25. Bell, J. P. and McCavill, W. T., J. Appl. Polym. Sci. 1974, 18, 2243.
26. Dusek, K., Bleha, M., and Lunak, S., J. Polym. Sci. (Polym. Chem. Ed.) 1977, 15, 2393.
27. Morgan, R. J. and O'Neal, J. E., Polym. Plast. Tech. Eng. 1978, 10, 49.
28. Schneider, N. S., Sprouse, J. F., Hagnauer, G. L., and Gillham, J. K., Polym. Eng. and Sci. 1979, 19, 304.
29. Morgan, R. J. and O'Neal, J. E., J. Mater. Sci. 1977, 12, 1966.
30. Morgan, R. J., Mones, E. T., and Steele, W. J., Polymer 1982, 23, 295.
31. Nielsen, L. E., J. Macromol. Sci. 1969, C3, 69.

RECEIVED January 7, 1983

12

Isothermal Cure Kinetics of an Epoxy Resin Prepreg

GARY L. HAGNAUER, BERNARD R. LaLIBERTE, and DAVID A. DUNN

Polymer Research Division, Army Materials and Mechanics Research Center, Watertown, MA 02172

The isothermal cure kinetics of a glass fiber-epoxy resin prepreg was investigated over the temperature range 80 to 135°C using high performance liquid chromatography (HPLC) and differential scanning calorimetry (DSC). Gel content was monitored gravimetrically. The prepreg resin (ca. 32% by weight) consisted of three different types of epoxy resins, an accelerator (Monuron) and the curing agent dicyandiamide. Cure kinetics parameters determined by DSC correlated with compositional parameters determined by HPLC. The curing chemistry of the prepreg resin was found to be highly dependent upon the cure temperature. Both the extent of reaction at the onset of gelation and crosslink density of the reaction products compared at the same % gelation increased with increasing temperature. The magnitude of the temperature effect suggests that curing temperature could have a profound effect on the network structure of the cured resin matrix and, therefore, on the properties of composites manufactured from this prepreg.

The isothermal cure kinetics of a glass fiber-epoxy resin prepreg is investigated using high performance liquid chromatography (HPLC) and differential scanning calorimetry (DSC). The prepreg has a recommended cure temperature of 121°C (250°F) and is used in the manufacture of structural composites. The prepreg resin (ca. 32% by weight) consists of three different types of epoxy resins, an accelerator (Monuron), and the curing agent dicyandiamide. HPLC is used to determine changes in the composition of the resin components and DSC is applied to monitor the overall extent of reaction as a function of cure time over the temperature range 80 to 135°C. Additionally, gel content is analyzed to relate cure kinetics to the formation and structure of the crosslinked epoxy resin matrix. The purpose of this work is to gain an understanding of the curing behavior of the prepreg and to determine whether variations in curing conditions may have a significant effect on the properties and reliability of the composite material.

Experimental

The prepreg consists of unidirectional S2 glass fibers (ca. 68% by weight) impregnated with a "staged" or partially reacted epoxy resin formulation. Because of possible batch-to-batch variations in prepreg composition, all the specimens for this study were selected from the same prepreg roll. In addition, experiments were conducted to ascertain the optimum specimen size for representative sampling and for determining resin composition. To prevent changes in composition due to aging, the prepreg was stored in a sealed container at $-14^{o}C$ and removed only for sampling.

The principal components in the resin were isolated by preparative liquid chromatography and identified using HPLC and spectroscopic techniques (1,2). The resin components are listed in Table I. Concentrations of the components in the as received "staged" prepreg resin and in a sample of the "unstaged" resin formulation are given. It is noted that the staged resin is not highly advanced. The actual extent of reaction based upon reacted epoxy and amine functionalities is probably no more than 1 or 2%. The epoxy equivalent weights in Table I were obtained from the manufacturer's technical literature. Monuron and dicy were assumed to have 1 and 4 equivalents per mole, respectively. The total epoxy and amine equivalents in the unstaged resin formulation are close to stoichiometric; i.e., 0.85 Eq. amine per 1 Eq. epoxide.

Partially cured prepregs for HPLC analysis were obtained by heating samples (1.7g) for varying periods of time in a thermostated oven. Immediately after the samples were removed and cooled, solutions were prepared by extracting the weighed prepreg with THF. Solution concentrations (ca. 10μg/μl) were calculated from the difference between the initial weight of the prepreg and the final weight of the fibers after extraction, drying, and treatment in a muffle furnace at $700-800^{o}C$. The weight percentage of insoluble reaction products (gel content) was calculated from the weight of the non-extractable organic material on the glass fibers. The extractable resin was essentially fully soluble. The solutions were prepared in volumetric flasks and were filtered through 0.2μM Millipore membrane filters.

A Waters Associates ALC/GPC-244 instrument with 6000A solvent delivery system, 660 solvent programmer, 710B WISP auto-injection system and a Perkin Elmer LC75 variable UV absorbance detector was used for the HPLC analyses. All analyses were run using a Waters μBondapak C_{18} (30cm x 3.4mm ID) column and a Spectra Physics SP4000 data system for peak integration and data formatting. Reagent grade water was prepared from distilled water using a Milipore Milli-Q2 water purification system. Distilled-in-glass, UV grade tetrahydrofuran (THF) was used as received from Burdick & Jackson Labs. Sample solutions were analyzed using the following conditions: injection volume -

Table I.
HPLC ANALYSIS OF PREPREG RESIN

	Component	Equiv. Weight (g/eq.)	Weight-% in Staged Resin	Weight-% in Unstaged Resin	Eq./100g in Unstaged Resin
1	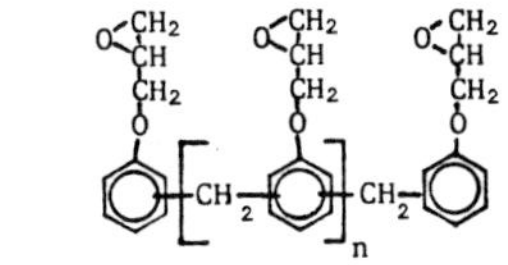 ECN 1273	225	41.9	46.3	0.206
2	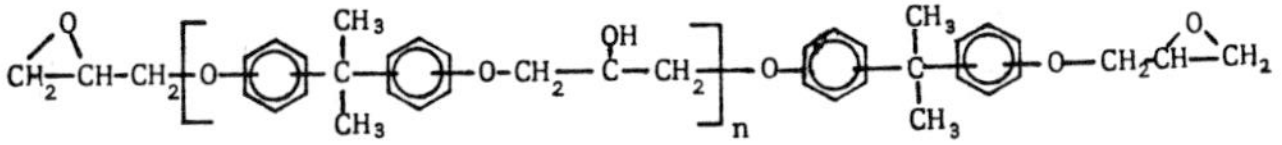 Epon 828	185	37.6	38.2	0.206
3	Reactive Epoxy Diluent	490	ca 5.0	ca 5.0	0.010
4	$H_2N-C(=NH)-NH-C\equiv N$ Dicyandiamide (dicy)	21	6.8	7.1	0.338
5	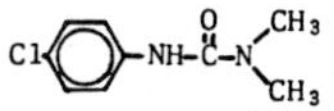 3-(p-chlorophenyl)-1-,1-dimethylurea (Monuron)	198.5	4.34	4.34	0.022

20 µl; mobile phase - (40% THF/60% H_2O) to (80% THF/20% H_2O) using gradient 8 and run 50 min.; flow rate - 2ml/min; detector - UV 237nm. A standard solution of the resin formulation was used for calibration.

Typical chromatograms for the unstaged resin formulation and a partially cured prepreg resin are shown in Figure 1. At least one peak having a characteristic retention time is associated with each component in the resin. Isomers and oligomers of some of the epoxy components are noted. The concentration (weight %) of each component is directly proportional to the area(s) under its respective peak(s). Only the concentrations of the reactants are determined using this procedure. The reaction of one group on a multi-functional molecule results in a new species which either produces a new HPLC peak or becomes incorporated into the gel. As the curing reaction proceeds in the early stages, it will be assumed that the decrease in concentration of a multi-functional component is due to a reaction involving only one of its groups; i.e., the change in concentration is directly proportional to the extent of reaction. The detailed analysis of changes in the oligomer concentrations supports this assumption.

DSC measurements were made using a DuPont 990 thermal analyzer with 920 DSC module. Procedures described in the instrument operating manual were used for temperature and calorimetric calibration. The prepreg samples were weighed 65-70mg (ca. 21-23mg resin) in open aluminum pans and run in static air. For isothermal analyses, the instrument settings were adjusted to optimize data collection at each temperature. An initial baseline was established by equilibrating the instrument at a preset curing temperature with empty sample pans and a run was initiated by inserting a filled sample pan. Thermal equilibrium generally was regained within 2 min. after sample insertion and the exothermic reaction was considered complete when the recorder signal levelled off to a baseline. The baseline at the end of the reaction was extrapolated to determine the total area under the exotherm curve and hence the isothermal heat of cure q_{iso}. Dynamic DSC analyses were run at a heating rate of 2°C/min. between 30 and 200°C on each isothermally cured sample to obtain the residual heat of reaction q_{res}. Data from the isothermal analyses were digitized and evaluated using a Hewlett Packard 9830 computer. Sample weight losses were negligible over the temperature range investigated.

Results

Gel Content and HPLC Analysis. Examples of the changes in the prepreg resin composition which occur during isothermal cure are shown in Figures 2 and 3. The percentages are based upon the total weight of organic material in the prepreg specimens. Each data point represents a separate experiment. It is possible to estimate rate constants using this data; however, it is noted that the HPLC procedure determines only the concentrations of

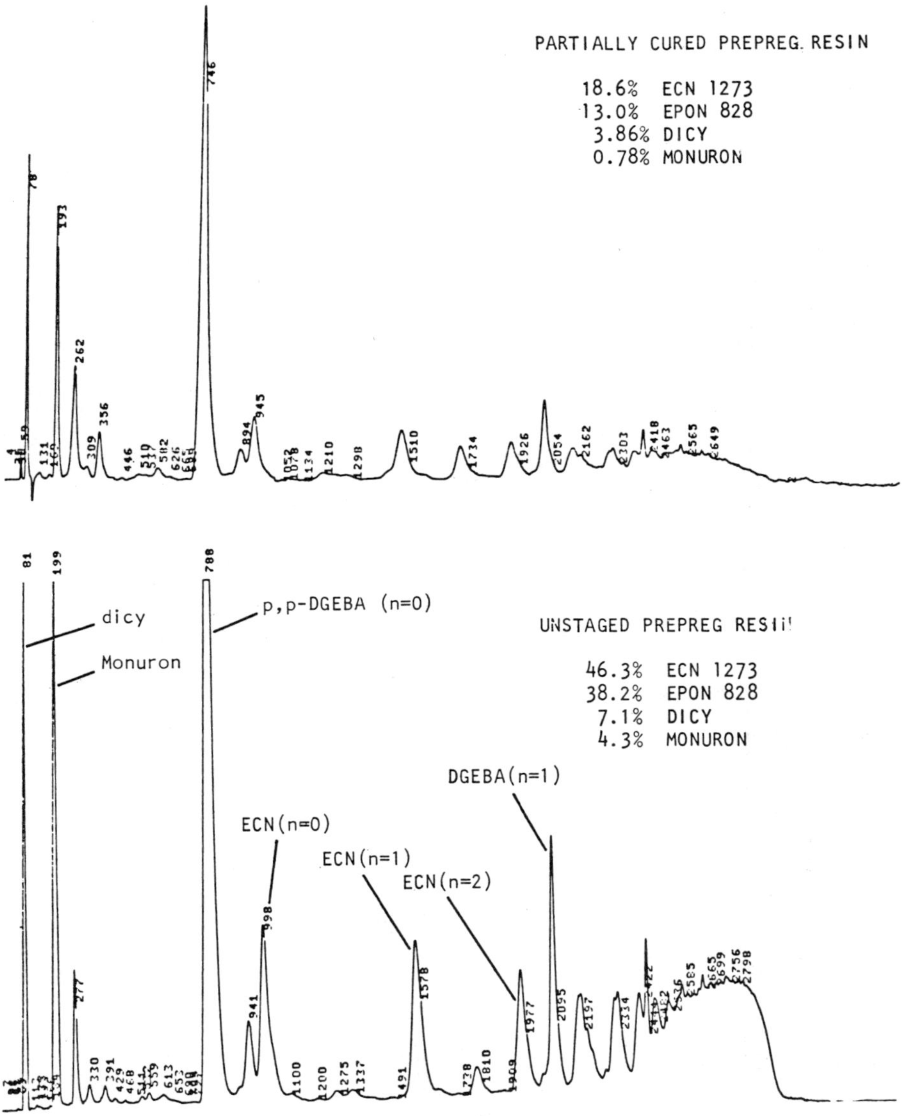

Figure 1. HPLC Analysis of Prepreg Resin.

unreacted molecular species and not the actual amine and epoxide concentrations needed for a thorough kinetics treatment. Therefore, no attempt will be made to determine kinetic constants from the HPLC data. Rather the HPLC and % gel data will be applied to ascertain whether or not the reaction pathway or

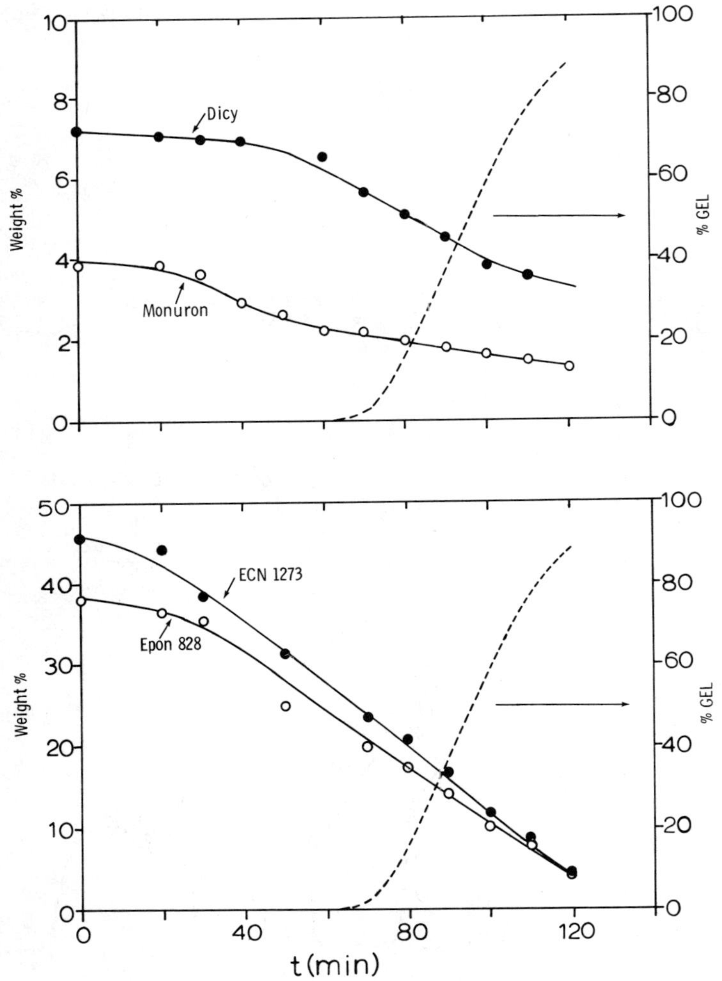

Figure 2. Changes in Prepreg Resin Composition During Isothermal Cure at 90°C.

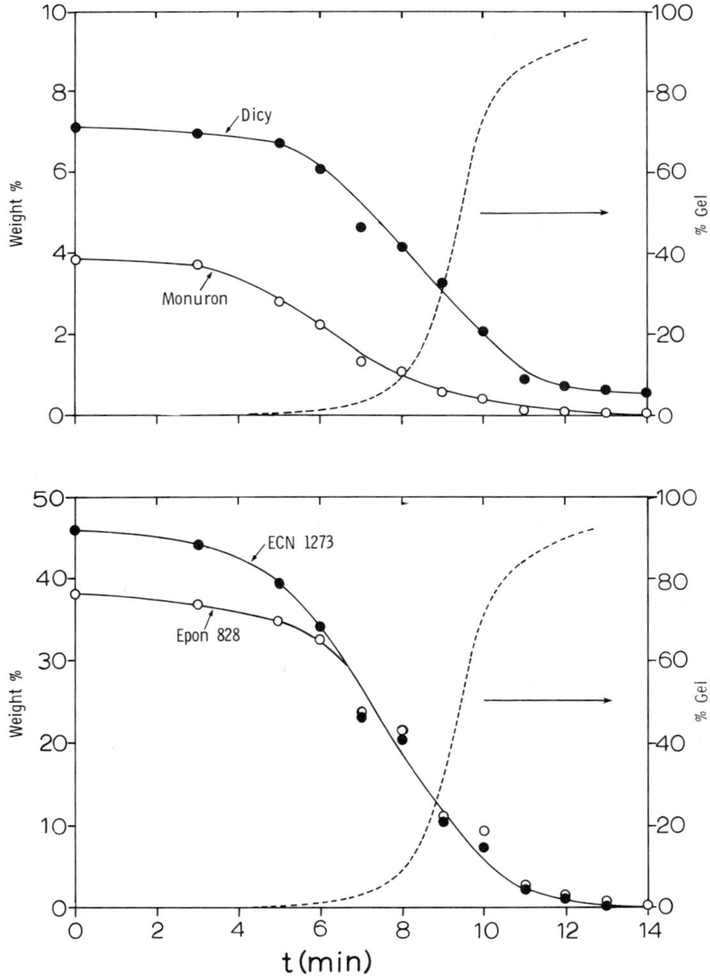

Figure 3. Changes in Prepreg Resin Composition During Isothermal Cure at 120°C.

mechanism of the epoxy resin polymerization depends on the isothermal cure temperature.

Comparing Figures 2 and 3, it is noted that at 30% gelation the weight percentages of unreacted ECN 1273, Epon 828, dicy and Monuron are respectively 17.5, 15, 4.6 and 1.8% for the 90°C cure and 11, 11, 3.0 and 0.6% for the 120°C cure. An additional 14.2% ECN 1273, 10.5% Epon 828, 22.5% dicy and 30% Monuron reacts at 120°C for the curing reaction to produce 30% gel. From this result it is concluded that the network structures of products formed at 120°C are more highly crosslinked than those formed at 90°C, at least in the early stages of gelation.

To further illustrate the effect of cure temperature on the reaction pathway, the ratio of the moles of dicy reacted to the average number of moles of epoxy-containing species reacted is plotted versus curing time at different temperatures (Figure 4). The moles of dicy and epoxy reacted are determined by subtracting the moles of each unreacted component from their respective initial mole concentrations. Hence only a single amine proton on a dicy molecule or a single epoxide group on an epoxy molecule is needed to react for the molecule to be considered part of the reaction product. Dashed lines are drawn in Figure 4 connecting data at the onset of gelation and at 50% gelation. Between 90 and 130°C the mole ratio at both the onset of gelation and 50% gelation nearly doubles. This data indicates the magnitude of the temperature effect and suggests that the effect persists to relatively high extents of reaction and hence that the curing temperature may have a profound effect on the network structure of the cured resin matrix.

DSC Analysis. A number of prepreg samples were partially cured using the DSC instrument at selected temperatures and analyzed by HPLC. The ratios of the peak areas of the resin components were compared with those obtained from the analysis of larger samples (1.7g) cured in an oven under the same conditions and were found to be in excellent agreement. Therefore, it was concluded that the curing behavior (rate of reaction and gel formation) of samples run in the DSC was comparable to that of the larger samples used to determine compositional changes during isothermal cure and discussed in the previous section.

The total heat of reaction q_t calculated from the sum of the isothermal and residual heats, $q_{iso} + q_{res}$, ranged from 60 to 85cal/g (resin) but showed no dependence on reaction temperature. However, the % residual heat decreased from 33% at 80°C to less than 1% for samples cured above 130°C (Figure 5). Also, in determining q_{res} it was noted that the temperature of the exotherm peak maximum T_{max} increased with increasing isothermal cure temperature (Figure 5). The incomplete reaction of the isothermal cures and the increase in the dynamic DSC exotherm temperature are most likely related to changes in the glass transition temperature T_g. As the resin cures and the T_g rises above the isothermal cure temperature, diffusion takes control

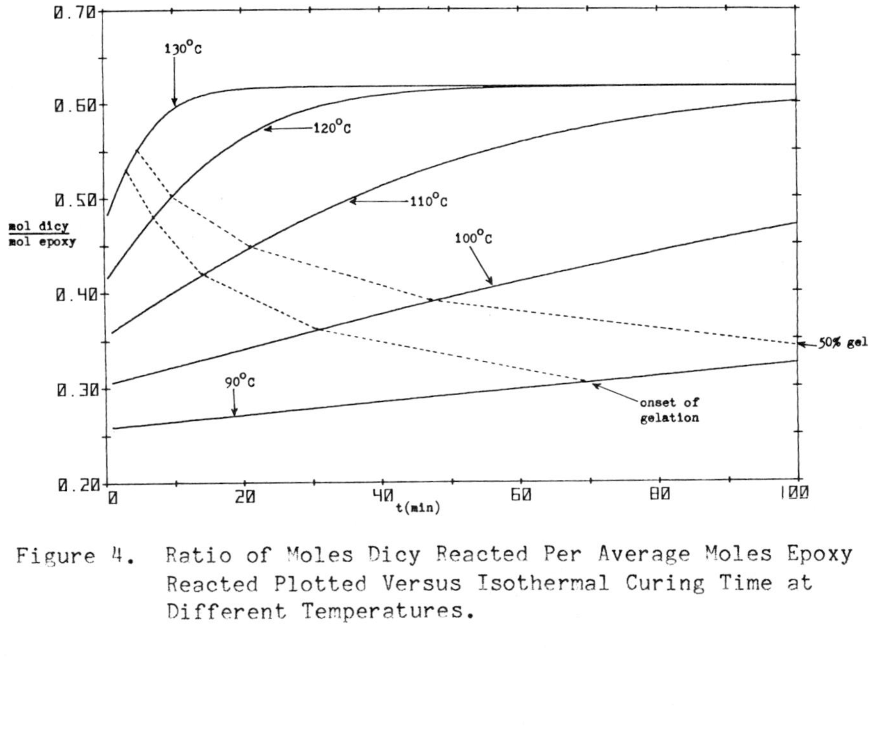

Figure 4. Ratio of Moles Dicy Reacted Per Average Moles Epoxy Reacted Plotted Versus Isothermal Curing Time at Different Temperatures.

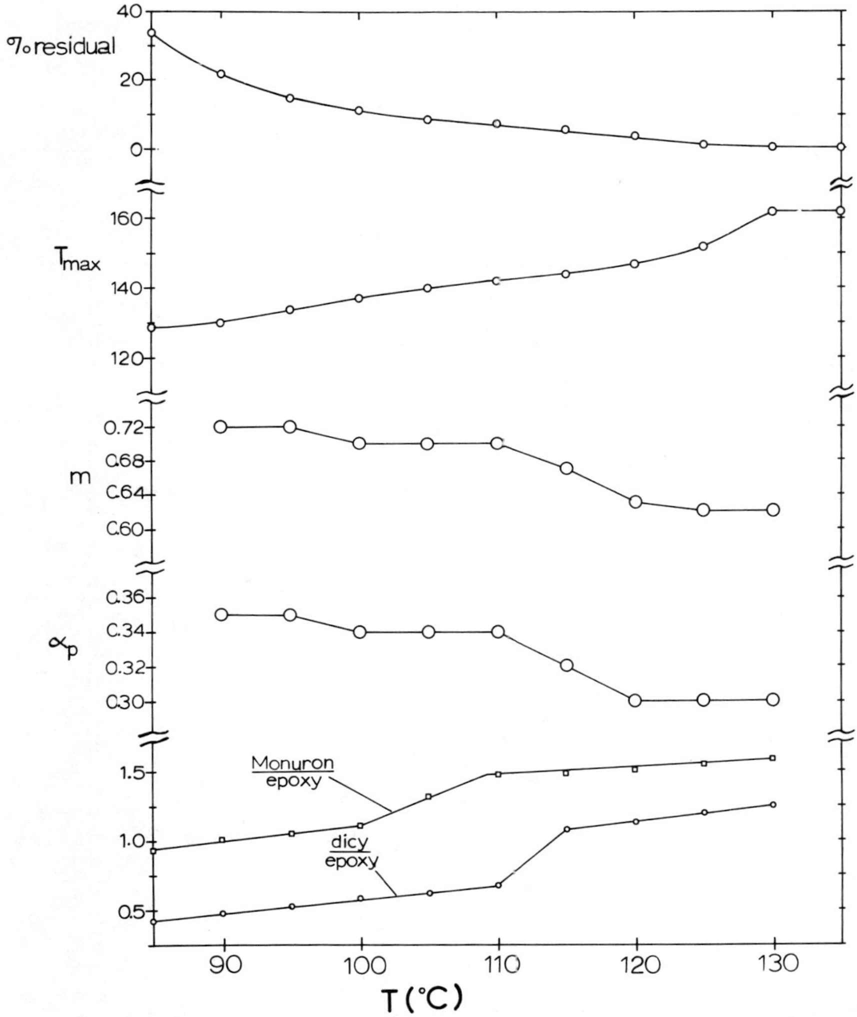

Figure 5. Extent of Reaction (α) Versus the Logarithm of the Isothermal Curing Time at Different Temperatures.

and curing eventually ceases before all the functional groups have reacted.

The difference in q_t values is attributed to variations in resin content due to the small size of the DSC samples. Therefore, it was necessary to determine q_t for each sample in order to obtain accurate extents of reaction α_i at time t_i for isothermal cures; i.e.,

$$\alpha_i = \sum_{j=1}^{i} q_j/q_t \tag{1}$$

where $\sum_{j=1}^{i}$ is the cumulative heat generated or area integrated under the DSC exotherm over the time interval $t_1=0$ to t_i and q_t is the total heat of reaction or integrated area. The extent of reaction is assumed to be proportional to the heat evolved during cure. The time/temperature dependence of the isothermal cures is illustrated in Figure 6.

The isothermal cure kinetics were evaluated using the following equation (3)

$$\dot{\alpha}=d\alpha/dt = (K_1+K_2\alpha^m)\ (1-\alpha)^n \tag{2}$$

where $\dot{\alpha}$ is the rate of reaction, K_1 and K_2 are rate constants, and m and n are exponents related to the order of the reaction. This expression describes the autocatalytic character of the epoxy cure where stoichiometric amounts of amines and epoxies are present. Furthermore, it is assumed that the amine protons are equally reactive. The kinetic parameters were calculated according to the method of Ryan and Dutta (4) in which the reaction is considered to be second order, $\overline{m+n}=2$, and the peak of the isothermal curve is defined by

$$\frac{d^2\alpha}{dt^2} = 0 \tag{3}$$

The constant K_1 was determined at $\alpha=0$ by extrapolation

$$K_1 = (d\alpha/dt)_{\alpha\ 0} \tag{4}$$

and K_2 and m were calculated

$$K_2 = (2-m)K_1 \cdot \alpha_\rho^{1-m}/(m-2\alpha_\rho) \tag{5}$$

$$m = \ln\left(\frac{\dfrac{\dot{\alpha}_\rho}{(1-\alpha_\rho)^{2-m}} - K_1}{\dfrac{(2-m)K_1\alpha_\rho^{1-m}}{m-2\alpha_\rho}}\right)\Big/\ln\alpha_\rho \tag{6}$$

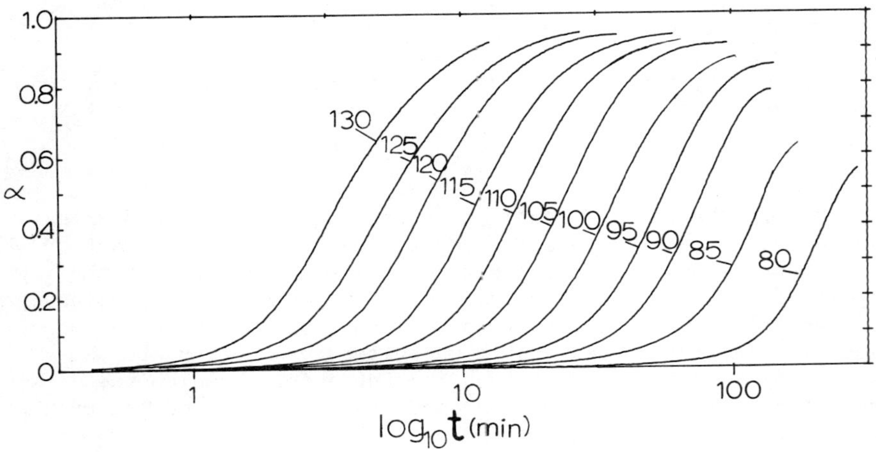

Figure 6. Rate of Reaction ($\dot{\alpha}$) Versus Extent of Reaction (α) at Different Curing Temperatures.

from K_1 and the rate $\dot{\alpha}_p$ and extent of reaction α_p at the peak of reaction rate curve α vs $\dot{\alpha}$ (Figure 7).

Significant results are illustrated in Figures 5 and 7. As the isothermal curing temperature is raised, the onset of gelation takes place at higher extents of reaction and the formation of an insoluble gel network occurs over a narrower curing range. The extent of reaction at the exotherm peak α_p decreases from 0.35 at 90°C to 0.30 at 130°C and exhibits a marked change between 110° and 120°C. The kinetic exponent m undergoes a similar change. In Figure 5, the ratios of the mole % of Monuron and dicy reacted to the mole % of epoxy components reacted at the onset of gelation is plotted vs the isothermal curing temperature. Both ratios increase with temperature and show a large change (transition) over a relatively narrow temperature range. The sharp increase in the reaction of dicy corresponds to the decrease noted for α_p and m between 110° and 120°C.

The rate of reaction $\dot{\alpha}_p$ and time t_p to the exotherm peak and the calculated rate constants K_1 and K_2 follow an Arrhenius dependence on temperature over the range 90–130°C (Figure 8). The activation energies are 23.8, 22.1, 23.2 and 20.9 Kcal/mol for $\dot{\alpha}_p$, t_p, K_1 and K_2, respectively, and are in excellent agreement with the value 22.0 Kcal/mol calculated from the time to the onset of gelation data. The data points for K_1 are connected with a curved line to illustrate the complexity of the temperature dependence of the curing reaction. The activation energy increases from 13.1 Kcal/mol between 90 to 100°C, to 27.7 Kcal/mol between 100 and 115°C, to 35.2 Kcal/mole between 120° and 130°C. The changes in activation energy relate to temperature intervals before, during and after the sharp increase in the ratios Monuron/epoxy and dicy/epoxy.

Conclusions

(1) The curing chemistry of the prepreg resin is highly dependent upon the cure temperature.

(2) The crosslink density of reaction products compared at the same % gelation increase with increasing cure temperature.

(3) Both HPLC and DSC show that the extent of reaction at the onset of gelation increases with increasing cure temperature.

(4) The extent of reaction over which an insoluble gel product forms becomes narrower with increasing cure temperature.

(5) There are correlations between isothermal cure kinetics parameters determined by DSC and compositional parameters determined by HPLC.

(6) DSC measurements are sensitive to changes in the curing chemistry of the prepreg resin.

(7) Precautions are advised when attempting to interpret the DSC measurements. Since the composition of the reaction product is temperature dependent, there are problems in defining α and comparing DSC data obtained at different temperatures.

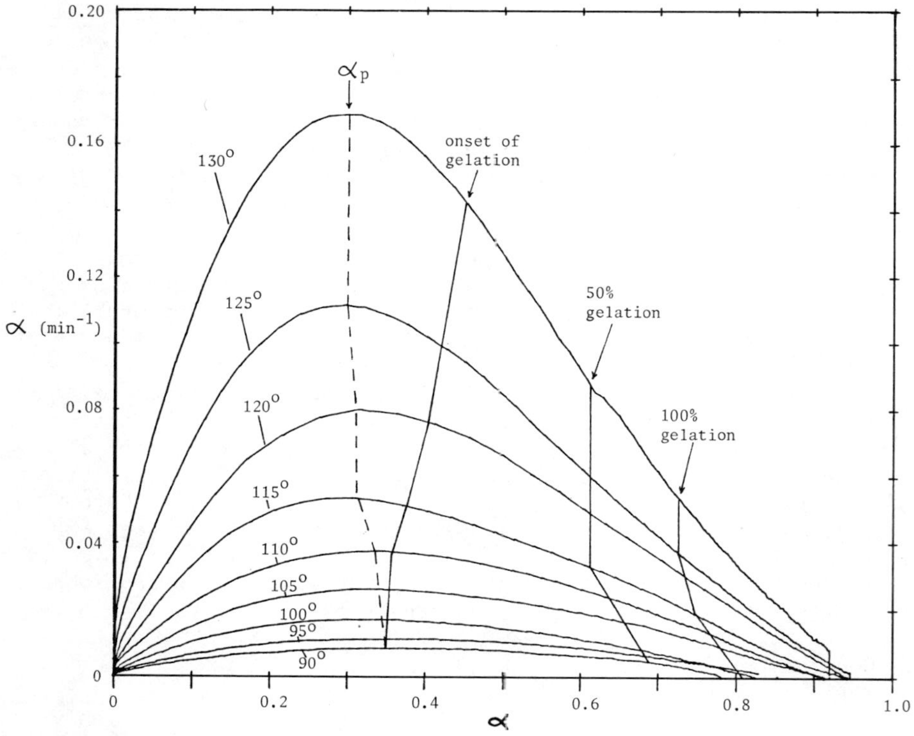

Figure 7. Comparative Plot. Temperature Dependence of the % Residual Reaction ($100\% \cdot q_{res}/q_t$), the Residual Exotherm Peak Maximum (T_{max}), the Kinetic Exponent (m), the Extent of Reaction at the Isothermal Peak Maximum (α_p) and the Ratios of Mole % Components Reacted.

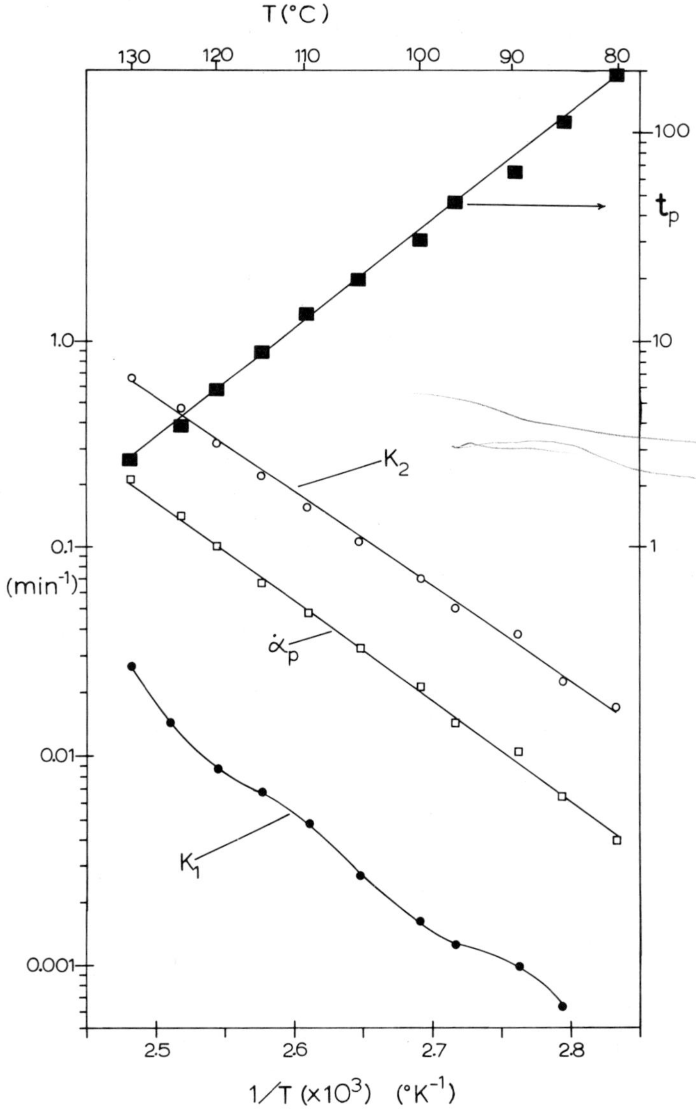

Figure 8. Arrhenius Plots. Temperature Dependence of the Isothermal Curing Time (t_p), the Rate of Reaction ($\dot{\alpha}_p$) at the Exotherm Maximum, and of the Kinetics Rate Constants (K_1 and K_2).

(8) The magnitude of the temperature effect on curing chemistry and the observation that the effect persists to relatively high extents of reaction suggests that curing temperature may have a profound effect on the network structure of the cured resin matrix and, therefore, on the properties of the composite. Likewise, variations in the cure cycle or heating rate could affect the reliability of the composite.

Comments

The isothermal cure kinetics of the prepreg has characteristics similar to those reported for the neat resin (5). Autocatalysis and diffusion control are observed and similar heats of reaction and activation energies are determined. The extents of reaction at the gel point estimated by DSC and torsional braid analysis of the neat resin(s) correspond to values determined gravimetrically in this study. The low value of the reaction ratio dicy/epoxy at the lower curing temperatures would explain the unreacted dicy noted by Schneider et. al (5).

Eventhough the prepreg has a complex, heterogeneous resin formulation and there are questions concerning the equivalency of the amine protons and the dicy curing mechanism (6,9), surprisingly consistent results are obtained using Equation 2 and assuming second order kinetics. The temperature dependence of α_o, m, K_1, dicy/epoxy and Monuron/epoxy indicates a complicated curing mechanism and suggests that the formation and structure of the crosslinked resin matrix are highly dependent on curing conditions. Hence variations in curing conditions may be expected to have a significant effect on the properties and reliability of the composite material.

Literature Cited

1. Hagnauer, G. L.; Setton, I. J. Liq. Chromatogr., 1, 55 (1978).
2. Hagnauer, G. L.; Dunn, D. A. "Materials 1980", 12th National SAMPE Technical Conference, Seattle, WA, 1980, Vol. 12, p. 648.
3. Sourour, S.; Kamal, M. R. Thermochimica Acta, 14, 41 (1976).
4. Ryan, M. E.; Dutta, A. Polymer, 20, 203 (1979).
5. Schneider, N. S.; Sprouse, J. F.; Hagnauer, G. L.; Gillham, J. K. Polym. Eng. Sci., 19, 304 (1979).
6. Saunders, T. F.; Levy, M. F.; Seriono, J. F. J. Polym. Sci., A-1, 5, 1609 (1967).
7. Son, P.; Weber, C. D. J. Appl. Polym. Sci., 17, 1305 (1973).
8. Zahir, S. A. "Proc. 6th Intern. Conf. in Organic Coatings Science and Technology", July 14-18, 1980, Athens, Greece, p. 745.
9. LaLiberte, B. R.; Bornstein, J., Army Materials and Mechanics Research Center (1981), TR 81-34.

RECEIVED December 16, 1982

13

Branching of High Molecular Weight Polyhydroxyethers Based on Bisphenol-A

S. A. ZAHIR

Ciba-Geigy AG, Basel, Switzerland

S. BANTLE

Universität Freiburg, Institut für Makromolekulare Chemie, Federal Republic of Germany

The kinetics of the polyaddition reaction of diglycidyl ether of Bisphenol A with Bisphenol A in solution and in the melt has been studied using tetrabutyl ammonium hydroxide and benzyl triethyl ammonium chloride as catalysts at several temperatures. The weight average molecular weight (M_w) and the number average molecular weight (M_n) can be related to the extents of reaction of the phenolic hydroxyl groups (α), the epoxide groups (β) and the branching probability p using a statistical method based on the cascade theory of branching processes. p is defined as the number of branches per secondary hydroxyl group in the chain. From p a kinetic parameter, $b = k_2/k_1$, can be determined, where k_2 is the rate constant for the addition of epoxide to the secondary hydroxyl groups in the chain (the chain branching step), and k_1 the rate constant for the addition of epoxide to the phenolic hydroxyl groups (the chain lengthening step). From the kinetic and molecular weight data, it has been possible to calculate b. The application of the branching theory and the measurements of the kinetics of the reaction are of considerable practical importance, as it enables the rational selection of catalysts, and allows the prediction of suitable reaction conditions for the synthesis of high molecular weight phenoxy resins.

One of the methods for the synthesis of high molecular weight polyhydroxy ethers based on Bisphenol A is the base catalyzed polyaddition of the diglycidyl ether of Bisphenol A (BADGE) with Bisphenol A (BPA) in the melt or in solution (Scheme 1).

0097–6156/83/0221–0245$06.00/0

Scheme 1 (Chain lengthening step)

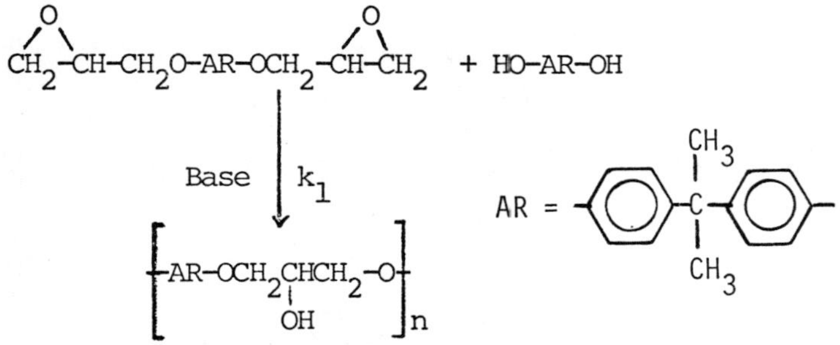

where k_1 = the rate constant for the addition of the epoxide group to the phenolic hydroxyl group.

The final molecular weight is normally controlled by the reaction stoichiometry and the purity of the reagents. During the chain lengthening step secondary hydroxyl groups are formed. The possibility of chain branching caused by the addition of the epoxide to the secondary hydroxyl group formed during the chain lengthening step leading to chain branching cannot therefore be altogether ignored (Scheme 2).

Scheme 2 (Chain branching step)

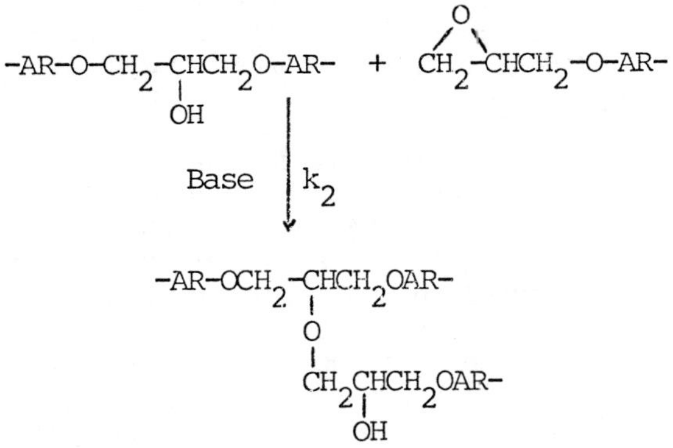

where k_2 = the rate constant for the addition of the epoxide group to the secondary hydroxyl group.

The extent of the latter reaction is a function not only of the nature and concentration of the catalyst used, but also of the temperature at which the melt (or solution) polyaddition reaction is carried out. Extensive branching can lead even at low conversions of the phenolic OH group to gelation. It is, therefore, of considerable practical importance to determine the extent of branching in such systems. Earlier NMR studies (1)on conventional epoxide resins (M_n = 1500-4000) has shown that the branching points per molecule varies between 0.09 to 0.6.

In a previous paper (2) we have shown by a statistical method based on the cascade theory of branching processes that closed analytic expressions can be derived for the weight average (M_w) and number average (M_n) molecular weights in terms of the mole ratio $R = m_b/m_a$ and the extents of the reaction of the phenolic (α), and the epoxide (β) functional groups, and the branching probability p, i.e., the probability that an epoxide group has reacted with a secondary hydroxyl group. m_b and m_a are the mole fractions of the diglycidyl ether of Bisphenol and Bisphenol A respectively. Using a simple kinetic model (Schemes 1 and 2), it can be shown that α, β, and p are related to a kinetic parameter $b = k_2/k_1$, where k_1 and k_2 are the rate constants for the chain lengthening and chain branching steps respectively (Eq.1).

$$b = k_2/k_1 = \left[\alpha\, p/(1-p)\right] \Big/ \left[-\ln(1-\alpha) - \alpha\right] \qquad (1)$$

$$\alpha = \beta\, R\, (1-p) \qquad (2)$$

When extensive branching occurs the system can gel. This occurs when $M_w \to \infty$. Under these conditions the critical branching probability , p, is related to the extent of reaction of the phenolic groups at the gel point, α_c, by

$$p_c = \frac{(R + \alpha_c) - (\alpha_c^4 + \alpha_c^2 + 2\,\alpha_c R)^{1/2}}{R + \alpha_c^2} \qquad (3)$$

The number average molecular weight is related to p by the equation,

$$M_n = \left[M_a + R.M_b\right] \Big/ \left[1 + R - 2\,\alpha/(1-p)\right] \qquad (4)$$

where M_a and M_b are the molecular weigths of bisphenol A (228.29) and the diglycidyl ether of bisphenol A (340.42) respectively.

The increase in the weight average molecular weight, M_w as a function of α, the extent of reaction of the phenolic groups, for different values of the branching parameter $b = k_2/k_1$ is illustrated in Figure 1. For b = 0 there is no addition of epoxide groups to the secondary glyceryl OH groups formed during the chain lengthening step. A linear polymer is formed with a pronounced increase in M_w near $\alpha = 1$. However, when $b > 0$ there is a finite probability of branching. From Figure 1 it is seen that gelation occurs at $\alpha_c = 0.93$ for b = 0.02, and at $\alpha_c = 0.815$ for b = 0.10.

Table I illustrates how the number average molecular weight at the gel point (M_n^g) varies as a function of the extent of reaction of the phenolic OH groups at the gel point, α_c, and b. It is seen that M_n^g is finite at the gel point and only shows a pronounced increase at high levels of conversion, and at very low levels of b.

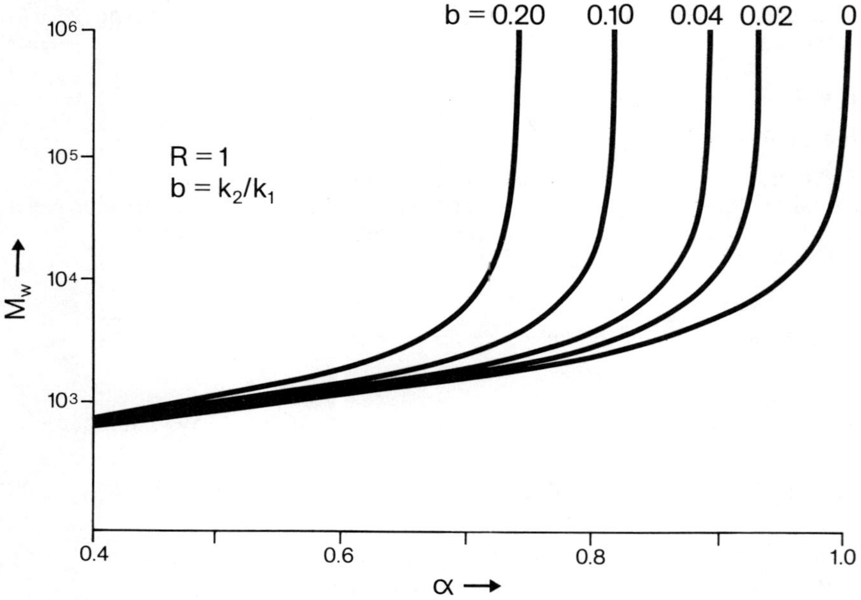

Figure 1.
M_w vs α for R = 1.00 and different values of b.

Table I. Number average molecular weight at gelation as a function of α_c for R = 1.00

α_c	β_c	M_n^g	b
0.95	0.97	11200	12.1×10^{-3}
0.96	0.98	14100	8.76×10^{-3}
0.97	0.985	18800	5.87×10^{-3}
0.98	0.99	28300	3.39×10^{-3}
0.99	0.995	56750	1.38×10^{-3}
0.9925	0.9962	75600	0.96×10^{-3}
0.995	0.9975	114000	0.58×10^{-3}
0.9975	0.9987	227000	0.25×10^{-3}

From equations 3 and 1 it is also possible to calculate the b values necessary to cause gelation in the synthesis of low molecular weight epoxide resins by the polyaddition reaction. For example, for a fixed value of $\alpha_c = 0.99$, b values are calculated for different R's, the ratio of epoxide to bisphenol A groups. The results are summarized in Table II. Thus in the synthesis of an epoxide resin with a calculated epoxide value of 0.48 (R = 1.15), gelation would be predicted to occur when the reaction is driven to 99% conversion of phenolic OH groups when a catalyst is used which promotes the addition of the epoxide groups to the aliphatic hydroxyl groups at a rate equal to 1.2% of the rate of the addition of the epoxide groups to the phenolic groups (b = 0.0116).

Now β can be measured by a titrimetric method and α by a spectroscopic method. Hence from Eqs. 1 and 2, p and b can be calculated. If M_w is also measured as a function of α, extrapolation of the plot $1/M_w$ vs α to $1/M_w = 0$ i.e. $M_w \rightarrow \infty$, leads to the critical conversion or extent of reaction of phenolic OH groups at the gel point, α_c. Thus b can be estimated from M_w or melt viscosity measurements near the gel point.

In this paper we report experimental results on the kinetics of the polyaddition reaction in the melt and in solution (diglyme) using two different catalysts, namely tetrabutyl ammonium hydroxide (TBAH) and benzyl triethyl ammonium chloride (BTAC) at different temperatures. By simultaneous measurements of α, β and M_w in two cases we have also evaluated the critical branching parameters.

Experimental

a) Chemicals: Bisphenol A (BPA) used was Fluka purum (m.p. = 154-156°C). The diglycidyl ether of Bisphenol A was purified from a commercial resin, ARALDITE® GY 250 (epoxide value = 5.25 eq/kg), first by molecular distillation, and then twice subjected to a crystallization and washing procedure in methanol. The final product (BADGE) had a melting point of 42-44°C, epoxide value = 5.88 eq/kg, hydrolysable chlorine = 0,006 eq/kg, and total chlorine = 0.01 val/kg, free OH< 0.1 eq/kg. Benzyl triethyl ammonium chloride was obtained from Fluka (purum), m.p. = 185°C (dec). It was used as a 0.1 N solution in isopropanol. Tetrabutyl ammonium hydroxide (Fluka) - (40% aqueous solution) was diluted in diglyme to give a 0.4 N solution. Dimethyl ether of diethylene glycol (diglyme) was of Fluka puriss quality (b.p. 60-62°/11mm, $n_D^{20} = 1.408$). Tetrahydrofuran (THF): THF (Fluka) of puriss quality was distilled over NaOH pellets to free it from inhibitor.

b) Analytical: Phenolic hydroxyl groups were determined by a modified spectroscopic method (4). The method was based on the conversion of the phenolic hydroxyl group to the phenolate anion by the addition of a standard aqueous solution

Table II. b as a function of R for a fixed value at gelation of $\alpha_c = 0.99$

R	Epoxide (val/kg)		phenolic OH (val/kg)	b
	calc.	at gel point		
1.00	0.00	0.018	0.035	1.38×10^{-3}
1.05	0.17	0.13	0.034	4.82×10^{-3}
1.10	0.33	0.27	0.033	8.21×10^{-3}
1.15	0.48	0.38	0.032	11.56×10^{-3}
1.20	0.63	0.49	0.031	14.88×10^{-3}
1.35	1.02	0.79	0.029	24.60×10^{-3}
1.50	1.35	1.05	0.027	34.01×10^{-3}

of tetrabutyl ammonium hydroxide. A difference spectrum (λ = 260-360 nm, λ_{max} = 304-305 nm) of the neutral polymer solution in THF/CH_3OH was compared with that of the same solution containing 20 μl of 0.5N TBAH in 1 cm quartz cells. Bisphenol A was used as the calibration standard. Depending on the solubility of the polymer, 3, 4 or 5 volume % of methanol in THF was necessary to obtain a clear solution. The extinction coefficients were $\epsilon_{3\%\ CH_3OH}$ = 2987; $\epsilon_{4\%\ CH_3OH}$ = 2963; $\epsilon_{5\%\ CH_3OH}$ = 2908 $l.eq^{-1}\ cm^{-1}$.

Epoxide values were determined by a well established potentiometric titrimetric method using perchloric acid in acetic acid (3). Weight average molecular weights were measured in a SOFICA light scattering apparatus in diglyme.

Melt-polyaddition: BADGE was heated in a stirred reaction vessel to 50°C. The bisphenol A was added portionwise and when a clear solution was obtained the mixture was brought to 20°C below the desired reaction temperature. A small sample was taken to measure the starting epoxide and phenolic OH values. The required catalyst solution was then injected into the melt, and the melt brought quickly to the desired temperature. Sampling was begun as soon as the temperature was reached and continued at regular intervals until a point was reached, when because of the massive increase in viscosity, stirring was difficult. The reaction mixture was poured into thin Al pans and placed in an oven maintained at the polymerization temperature. Sampling was then continued. The temperature was maintained at ± 1°C.

Solution polyaddition: To the correct proportions of BADGE and BPA sufficient diglyme was added to prepare a 50% (w/w) diglyme solution. The solution was brought to the desired temperature, a reference sample taken, and the catalyst solution in diglyme injected to start the reaction. Samples were taken at regular intervals.

Results

Tetrabutyl ammonium hydroxide (TBAH) as catalyst. Figure 2 shows the plot of α vs β.R for the solution polyaddition at 100°C using TBAH as catalyst (R = 1.00, catalyst / bisphenol A = 0.002 eq/mol). There is only a slight deviation from a straight line indicating very little branching. An overall best fit for the experimental points were obtained for b = 0.005. The plot for the reciprocal weight average molecular weight ($1/M_w$) vs α for the same kinetic run is shown in Figure 3. Extrapolation to $M_w \rightarrow \infty$ gave a value of α_c = 0.985 at gelation. Substituting this value of α_c in equations 3 and 1, a value for the branching parameter b = 0.0028 was obtained. When the same kinetic run was repeated at 150°C significant amounts of branching was observed, as may be seen from Figure 4 (the plot of α vs β.R). A calculated value of b = 0.02 gave the best fit with the experimental data.

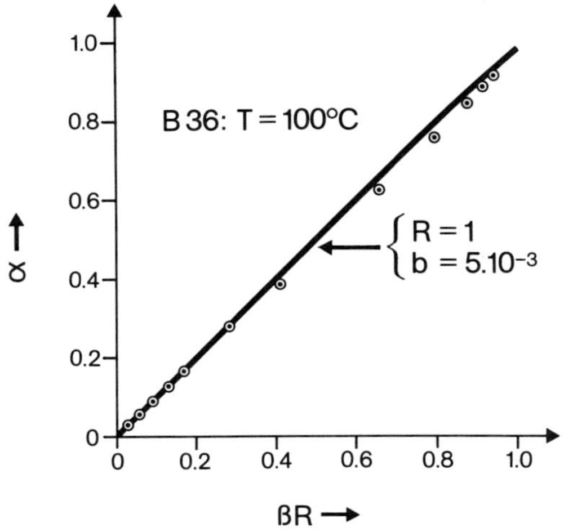

Figure 2.
α vs β.R at 100°C for catalyst TBAH (2.10^{-3} eq per mol BPA) and R = 1.00. The circles represent experimental values, and the full line the calculated curve for α vs β.R for b = 0.005.

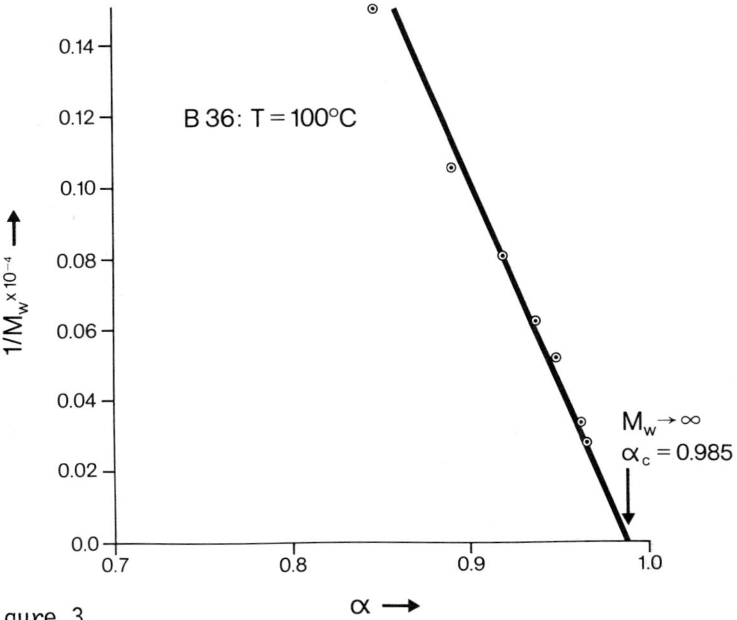

Figure 3.
$1/M_W$ vs α for catalyst TBAH (2.10^{-3} eq per mol BPA) at 100°C and R = 1.00. Extrapolation of the curve to $1/M_W$ = 0 gave an α_c = 0.985.

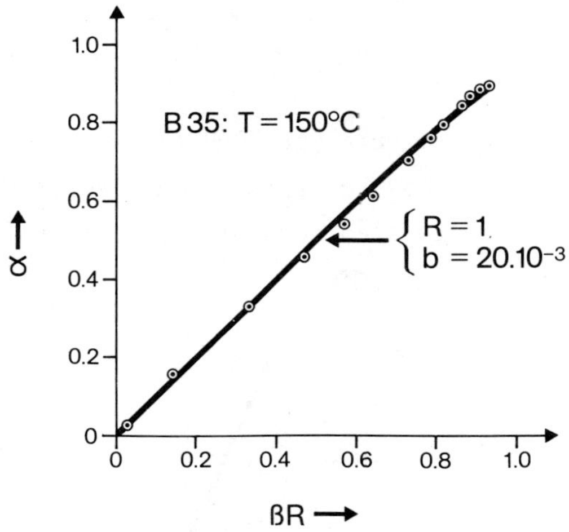

Figure 4.
α vs β.R at 150^{0}C for catalyst TBAH (2.10^{-3} eq per mol BPA) and R = 1.00. The circles represent experimental values and the full line the calculated curve α vs β.R for b = 0.02.

Benzyl triethyl ammonium chloride (BTAC) as catalyst. Preliminary experiments using BTAC as catalyst at 184°C showed that the reaction comes to a stop at a very early stage at 184°C but not at 150°C. These reactions were carried out in the melt, using a commercially available liquid epoxide resin, Araldite® GY 250, at a value of R = 1.00, and a catalyst/bisphenol-A ratio of 0.0001 mol/mol. The results are depicted in Figure 5. The reason for the lower degree of conversion at 184°C as opposed to 150°C is believed to be caused by the deactivation of the catalyst at 184°C. In Figures 6 and 7 the α vs β.R curves are plotted for the data depicted in Figure 5. Considerable amounts of branching are observed even at low degrees of conversion both at 150°C, (b = 0.5) and at 184°C (b = 0.9). The experiment with BTAC as catalyst was repeated at 180°C using the purified bisglycidyl ether of bisphenol A (BADGE) instead of the commercial resin Araldite® GY 250, and at a higher epoxide/phenolic-OH ratio, R = 1.2872. Again extensive branching was observed as shown by the α vs β.R plot, Figure 8. Best fit with the experimental values were obtained for a calculated value b = 1.00. From the $1/M_w$ vs α plot, Figure 9, for this latter run, α_c, the critical phenolic OH group concentration was found to be 0.513. Substitution into equations 3 and 1 of this value of α_c led to a b = 1.23.

Kinetics of the epoxide/phenolic OH reaction. At low extents of reaction the concentration of secondary glyceryl hydroxyl groups is low. The branching reaction can be ignored ($k_2 \rightarrow 0$), and the reaction of epoxide groups with phenolic OH groups can be assumed to follow second order kinetics (see Reference 2, Appendix II). It is thus possible to estimate k_1 the rate constant for the addition of epoxide to phenolic OH groups. Another useful parameter that can be deduced from the experimental data is the degree of branching, q (q =p/(1-p)). "q" is defined as the fraction of all repeat $-CH_2CHOHCH_2-$ units in the polymer that have reacted with epoxide groups to form branched ether linkages. The results from the kinetic and the molecular weight measurements for b, k_1,k_2 and q_c for the two catalysts used in this study, namely TBAH and BTAC, are summarized in Table III.

Discussion

Table III illustrates how important it is in practice to choose highly selective catalysts for the polyaddition reaction. With a strong base like TBAH as catalyst very little branching is observed at 100°C. At 150°C, however, k_2 is over a 100 times larger than at 100°C. Although this would normally have led to extensive branching, because of a nearly thirtyfold increase in the rate constant, k_1, for the epoxide/phenolic OH reaction branching is not very extensive.

The actual mechanism of the initiation of the polyaddition reaction by quaternary ammonium salts is probably very complex.

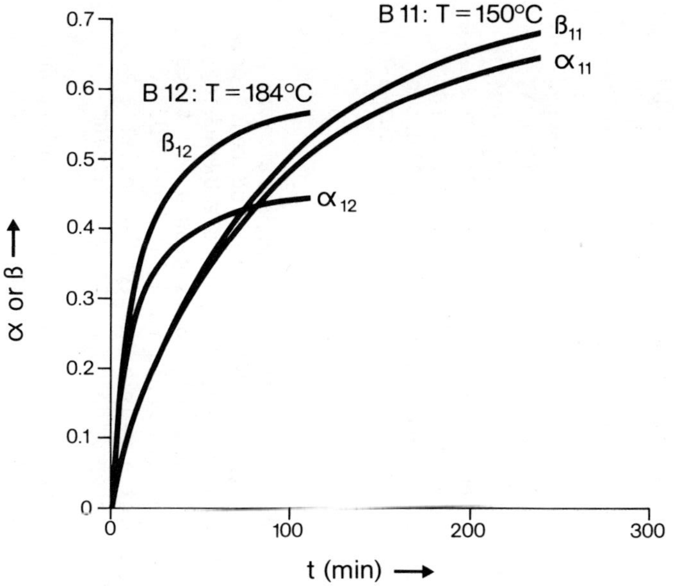

Figure 5. Kinetic curves of α and β as a function of reaction time for the catalyst BTAC at 150 °C and 184 °C. For the 150 °C run (System B11: α_{11}, β_{11}), [BATC] = 1.10^{-3} mol per mol BPA and R = 1.07). For the 184 °C run (System B_{12}: α_{12}, β_{12}), [BTAC] = 1.10^{-4} mol per mol BPA and R = 1.00. The resin used was ARALDITE GY 250, and the reaction was carried out in the melt.

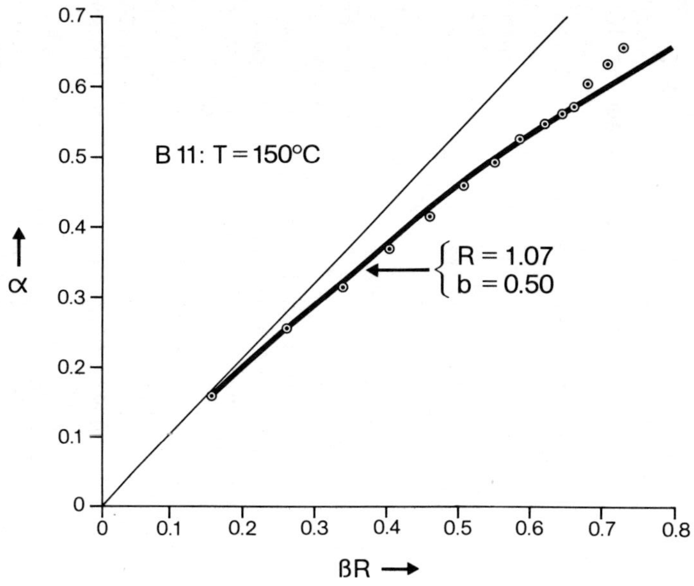

Figure 6. α vs β.R for run B_{11} (Catalyst = BTAC, [BTAC] = 1.10^{-3} mol per mol BPA, T = 150 °C, R = 1.07). The circles represent experimental values, and the full line the calculated curve α vs β.R for b = 0.5.

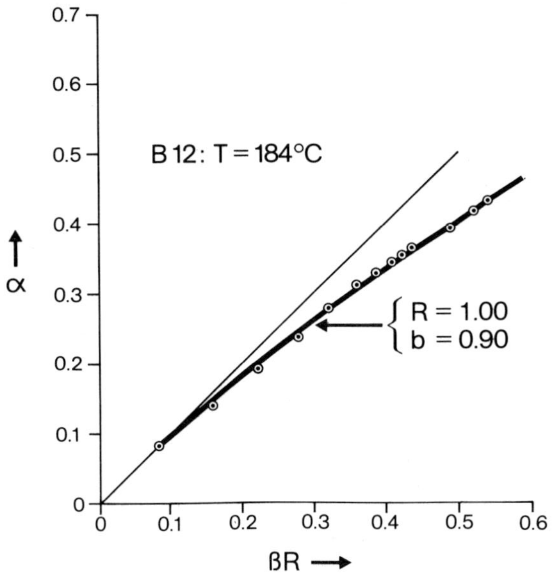

Figure 7. α vs β.R for run B_{12} (Catalyst = BTAC, [BTAC] = 1.10^{-4} mol per mol BPA, T = 184 °C, R = 1.00). The circles represent experimental values, and the full line the calculated curve for α vs β.R for b = 0.9.

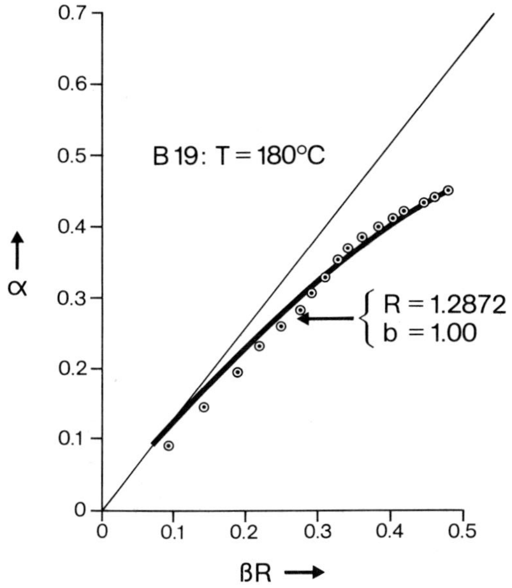

Figure 8. α vs β.R for run B_{19} (Catalyst = BTAC, [BTAC] = 1.10^{-4} mol per mol BPA, T = 180 °C, R = 1.2872). In this case pure bisglycidyl ether of bisphenol A (BADGE) was used as the epoxide resin. The circles represent experimental values, and the full line the calculated curve α vs β.R for b = 1.00.

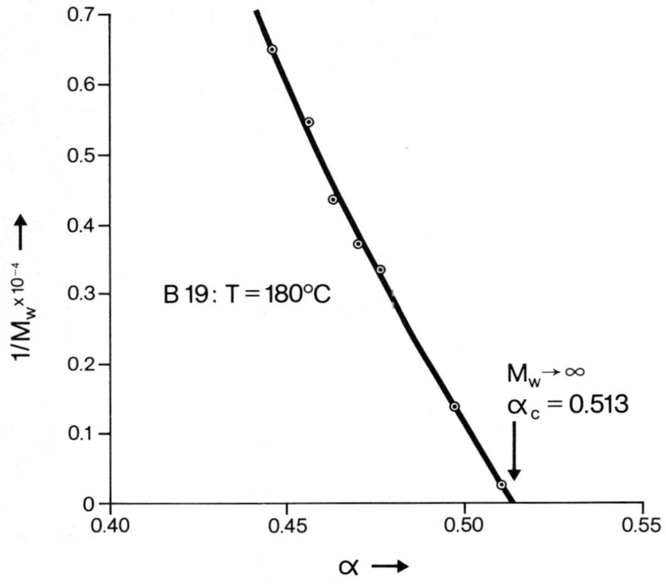

Figure 9.
The plot of $1/M_W$ vs α for run B19 (catalyst = BTAC, [BTAC] = 1.10^{-4} mol per mol BPA, T = $180^{o}C$, R = 1.2872, epoxide resin, BADGE). Extrapolation to $1/M_W$ = 0 led to α_c = 0.513.

Table III. Comparision of kinetic parameters, k_1, and k_2 for the catalyst systems TBAH and BTAC

EXPT	B 36	B 35	B 11	B 12	B 19
Catalyst	TBAH	TBAH	BTAC	BTAC	BTAC
Catalyst/BPA (mole/mole)	2.10^{-3}	2.10^{-3}	1.10^{-3}	1.10^{-4}	1.10^{-4}
Resin	BADGE	BADGE	GY 250	GY 250	BADGE
R	1.00	1.00	1.07	1.00	1.287
Solvent	Diglyme	Diglyme	Melt	Melt	Melt
Temperature °C	100	150	150	184	180
(From Kinetics)					
$b = k_2/k_1$	5.10^{-3}	20.10^{-3}	0.5	0.9	1.00
k_1 (kg.eq^{-1}. min^{-1})	$1.85.10^{-3}$	$54.3.10^{-3}$	$2.8.10^{-3}$		$1.4.10^{-3}$
k_2 (kg.eq^{-1}. min^{-1})	$0.009.10^{-3}$	$1.09.10^{-3}$	$1.4.10^{-3}$		$1.40.10^{-3}$
(From M_w)					
$b = k_2/k_1$	$2.77.10^{-3}$	-	-	-	1.23
q_c	0.009	-	-	-	0.5

It is interesting to note that the use of BTAC as a catalyst causes more extensive branching at low conversions than TBAH (compare Figure 6 with Figure 4). In the case of OH^- as nucleophile, as long as any phenolic OH is present, the concentration of free OH^- is very low. Only when the concentration of the phenolic OH drops to low levels, does the inherently faster reaction of the alkoxide ion with the epoxide takes over to cause branching. In the case of nucleophiles such as Cl^- or triethyl amine additional branching might also be caused by reactions leading to the formation of 1,3 dioxalanes, as illustrated in Scheme 3.

Scheme 3

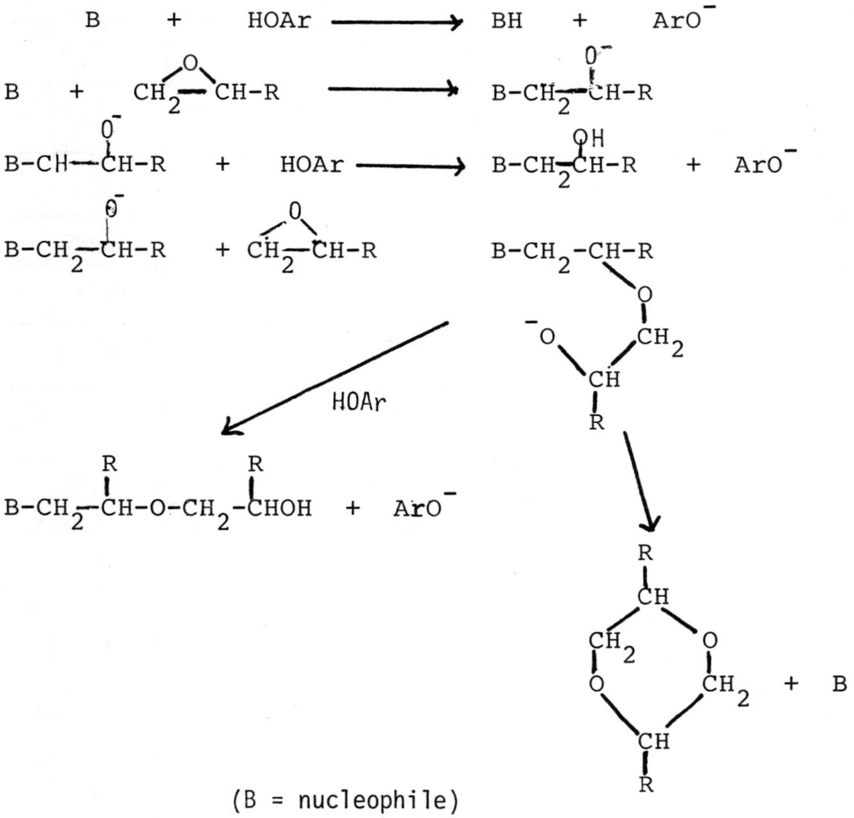

(B = nucleophile)

When B is triethyl amine, a quaternary ammonium salt is formed. Since triethyl amine is volatile at the temperatures usually used for polyaddition, the catalyst would be depleted from the reaction mixture as the reaction progresses. In the case of a catalyst such as benzyl triethyl ammonium chloride, which would be expected to undergo thermal decomposition into benzyl chloride and triethyl amine, such a mechanism would predict why not only extensive branching occurs, but also why the reaction comes to a complete stop when high temperatures are used for the polyaddition reaction.

Acknowledgment

We are much indebted to Professor W. Burchard of the Institut für Makromolekulare Chemie, Universität Freiburg, Germany, for extensive discussions. S.A.Z would like to express his thanks to the management of CIBA-GEIGY for permission to publish this paper.

Literature Cited

1. Batzer, H.,Zahir, S.A., J.Appl.Polym.Sci. 1975, 19, 601.
2. Burchard, W., Bantle, S., Zahir, S.A., Makromol.Chem. 1981, 182, 145.
3. Kerker,R., Störi, M., Chimia, 1979, 33, 84.
4. Schöri, E., McGrath, J.E., J.Appl.Polym.Sci. 1978, 34, 103.

RECEIVED December 2, 1982

14

Characterization and Moldability Analysis of Epoxy Reaction Injection Molding Resins

J. S. OSINSKI and L. T. MANZIONE

Bell Laboratories, Murray Hill, NJ 07974

The interest in extending the Reaction Injection Molding (RIM) process to non-urethane resins has generated need for a greater understanding of the molding behavior of these fast-cure systems. Epoxy, for example, is recognized as a candidate for commercial RIM, but its lower reaction rate and high heat of reaction render it more difficult to mold than polyurethane. The kinetics and rheology of such a resin must therefore be well characterized in order to undertstand the molding process and evaluate its feasibility. A method of fully characterizing an epoxy molding system by combining adiabatic reactor studies, infra-red spectrometry, and viscometer data is therefore described. This provides time-temperature-conversion-viscosity correlations which can be directly utilized in computer studies designed to minimize actual molding times while avoiding a damaging exotherm during cure. The particular system studied is a low viscosity liquid epoxy cured with aminoethylpiperazine and was found to be moldable over specific temperature ranges. Conducting an analysis of this type on a prospective resin can help determine the value of further investment in experimentation, material development, molding equipment, and tooling.

Reaction Injection Molding (RIM) is polymerization and polymer processing in a single cyclic operation. The process consists of metering and mixing low viscosity reactive fluids, mold filling, and polymerization or cure in the mold cavity. Although commercial RIM is primarily associated with polyurethane due to its wide use in the automotive industry, there is considerable interest and activity in extending the process to other resin systems. Epoxy, for example, has certain properties and characteristics which make it desirable for use in such an operation. Certain epoxies may be suitable for RIM since they are two-part,

0097-6156/83/0221-0263$06.00/0

low viscosity systems that provide a rapid, exothermic reaction that goes to completion without producing undesirable by-products.

However, epoxies are inherently less amenable to the RIM process than polyurethanes. Referring to Figure 1, we see obvious differences in the adiabiatic reaction paths of typical epoxy and polyurethane reactions. The epoxy reaction, starting at a higher temperature than the polyurethane, proceeds much more slowly but ultimately releases much more heat during the reaction. These differences will have important implications in the molding of the resin. The comparatively slower reaction rate of epoxies means that higher molding temperatures are required to achieve reasonable cycle times. It then becomes possible for the peak temperature achieved during cure to exceed the thermal degradation temperature of the system due to the high heat of reaction associated with epoxies. The understanding and prediction of such problems introduced by the use of epoxies will yield information on the processability and moldability of a given resin system in a particular application, obtainable without extensive experimentation. Most significant are the effects of mold and fluid temperatures, part thickness, and filler loading on the viscosity, cure time, and the maximum exotherm temperature.

Such an analysis, however, necessarily involves the acquisition and correlation of data related to the viscosity and kinetics of the curing polymer. Computer modeling of mold filling and curing events is then possible (1, 2, 3). Coupling these results with practical constraints such as ultimate physical properties of the cured resin, ease of handling, machine compatibility, and cost, a discriminating decision can hopefully be reached when comparing different resins, without actual molding trials of each candidate.

This chapter will describe techniques for evaluation of an epoxy RIM system from the above standpoints. First a description of the kinetic and viscosity characterizations will be presented, followed by a moldability analysis using these data.

Most experimental epoxy RIM systems consist of a diglycidyl ether of bisphenol A epoxy (DGEBA) with amine or anhydride curing agents (4). The two part liquid epoxy system chosen for the present analysis is a DGEBA-type resin plus reagent grade aminoethylpiperazine (AEP) as a curative. The epoxy has a molecular weight of 380 and an equivalent weight, based on reactive epoxy sites, of 190. AEP is a trifunctional amine of equivalent weight 43 containing primary, secondary, and tertiary amines, the latter of which is unreactive but acts as a catalytic site. This curing agent was chosen among the several that were considered because of its relative high rate of reaction and its potential for use as an actual off-the-shelf RIM component. All experiments were performed with the nominal stoichiometry of 22.6 parts by weight per hundred resin (phr).

Certainly faster systems are available than the one chosen

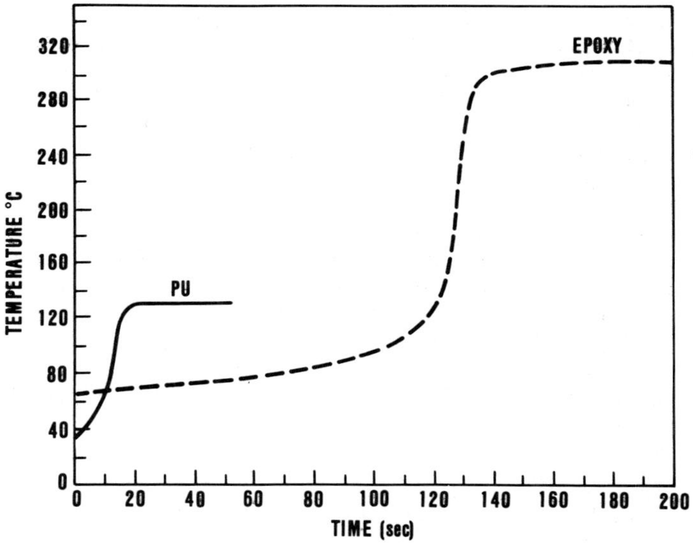

Figure 1.
Typical adiabatic temperature rise of epoxy and polyurethane reactions.

here, such as those using boron trifluoride complexes as a Lewis acid curing agent or those incorporating proprietary catalysts. However, based upon potential processability as a readily available off-the-shelf system, DGEBA epoxy+AEP looks most promising.

Resin Characterization

Theoretical treatment of this polymerization is difficult because of the presence of both primary and secondary amine reactions as well as tertiary amine catalyzed epoxy homopolymerization. To obtain kinetic and viscosity correlations, empirical methods were utilized. Various techniques that fully or partially characterize such a system by experimental means are described in the literature (5-9). These methods include measuring cure by differential scanning calorimetry, infra-red spectrometry, viscometry, and by monitoring electrical properties. The presence of multiple reaction mechanisms with different activation energies and reaction orders (10) makes accurate characterizations difficult, but such complexities should be quantified. A dual Arrhenius expression was adopted here for that purpose.

Adiabatic Reactor. The direct measurement of the adiabatic temperature rise of a reacting resin as a function of time was used as a means of generating kinetic data. This method has been applied to polyurethane by Lipshitz and Macosko (11), but a computerized adaptation of their method was found more useful for epoxies. Combining the results of this method with viscometer and IR studies, the effective epoxy kinetics of a given system are represented by:

$$-\frac{dC}{dt} = A_1 \exp\left|\frac{-E_{A_1}}{RT}\right| C^{n_1} + A_2 \exp\left|\frac{-E_{A_2}}{RT}\right| C^{n_2} \qquad (1)$$

(see List of Symbols). Thus, by correlating data over specific temperature regimes, the dominate activation energy and reaction order in that regime can be independently evaluated.

A typical epoxy reaction carried out adiabatically may have an exotherm well in excess of 200°C. Therefore, much of the reaction will be governed by mechanisms associated with high temperatures, and as a result predominated by kinetics different from those in the initial reaction region (under 125°C) (10). In such a case Equation 1 might effectively reduce to

$$-\frac{dC}{dt} = A \exp\left|\frac{-E_A}{RT}\right| C^n \qquad (2)$$

and we could obtain the high temperature parameters based on this simpler expression.

Weighed quantities of the epoxy resin and curing agent to be tested are heated separately in a convection oven to the desired initial temperature of approximately 50°C. Prior to data acquisition, the resin is transferred to a static oven held at the same initial temperature to minimize subsequent heat losses. Because high temperatures are reached, the reaction is carried out in an FEP Teflon beaker surrounded by a styrofoam jacket. A fast response (45 ms time constant) Chromel-Alumel thermocouple is submerged in the fluid and the signal is sent to a small amplifier and then input directly to a desktop-computer controlled data acquisition system. This system can accurately sample data points at intervals ranging from less than a millisecond to an indefinite time period. Generally the sampling rate used for the present system was ~1 per sec.

At this point a stoichiometric quantity of curing agent is quickly added to the container and vigorously stirred for approximately 15 sec as temperature data are obtained. Mixing can be carried out manually and is found to be effective. Heat losses near the end of the reaction are not entirely negligible and are accounted for in the subsequent data reduction routine as described below.

A heat balance of an adiabatic reaction based on Equation 2 yields an expression of the form:

$$\ln\left|\frac{\rho C_p(T)}{(-\Delta H)C_o(C/C_o)^n}\frac{dT}{dt}\right| = \ln A\, C_o^{n-1} - \frac{E_A}{RT} \tag{3}$$

Plotting the left side of this equation against 1/T should therefore yield a straight line, enabling calculation of the activation energy and pre-exponent from the slope and intercept, respectively. Material density is measured and assumed constant; this is reasonable because the small increase in density due to polymerization is partially offset by the simultaneous decrease due to high temperature. C_p as a function of temperature (see Table I) was obtained from an approximate polynomial fit to the data of Kamal and Ryan (6) for their epoxy system, which includes the effect of increasing chemical conversion. Heat losses are corrected for by determining the downward slope of the tail of the raw temperature vs. time curve and calculating the effective overall heat transfer coefficient. The heat loss at each time increment can then be calculated and added back into the system mathematically. ΔH and other values can be calculated from the adjusted data.

The derivative $\frac{dT}{dt}$ at each point is obtained by a fourth order central difference equation, then an arbitrary but approximate value of n is chosen and all points other than very early and plateau points are correlated by a least-squares linear regression using Equation 3. The index n is incremented or decremented systematically until the best correlation coefficient is obtained. Data from a single experimental run can be reduced alone or combined with other data files of the same system. Employment of

Table I
Reaction Kinetic Parameters

System	phr curative	C_o	ρ	$-\Delta H$	A	E_A/R	n	ΔT_{ad}	X_{gel}	r
AEP (Hi-temp)	22.6	4682	1090	129000	3.977	7991	2.80	233	-	.9987
AEP (Lo-temp)	22.6	4682	1090	-	.3603	4640	2.0	-	.49	.9996

No entry means not measured.

$C_p = -524 + 6.29T + .000686T^2$.

this routine on a test file generated by an nth order Arrhenius expression easily reproduces the parameters used, correct to four or more significant digits.

Sample outputs for epoxy + AEP are shown in Figures 2 and 3. Figure 2 illustrates the typical fit that is possible using data from a single experiment. Kinetic constants obtained from one trial do not satisfactorily predict other experiments at different starting temperatures, however. Therefore, it is advantageous to correlate several files of data simultaneously, as in Figure 3. The equation obtained when this is done represents the adiabatic reaction kinetics:

$$-\frac{dC}{dt} = 3.977 \exp\left|\frac{-7991}{T}\right| C^{2.80} \frac{\text{equiv.}}{m^3\text{-sec}} \tag{4}$$

Viscometry and Infrared Spectrometry. To obtain a low temperature kinetic expression as well as rheological data for epoxy + AEP, the viscosity of the curing resin as a function of time and temperature was obtained on a Ferranti-Shirley cone and plate viscometer and matched with conversion data acquired through IR spectrometry. No attempt was made to account for non-Newtonian effects since in practice mold filling should take place in the reasonably Newtonian region of low conversion. The material was found to be Newtonian over the range where cavity filling would be expected to take place. The shear rate used was approximately 150 sec^{-1}. Isothermal gel times at temperatures from 30 to 90°C at 10° intervals were found by extrapolating the asymptotic viscosity curve to infinite viscosity (see Figure 4), representing gelation. Assuming the conversion of reactive species at the gel point to be independent of the isothermal cure temperature, a plot of ln t_{gel} vs. 1/T should then give a straight line of slope $\frac{E_A}{R}$. This is indeed the case for the data obtained. Least squares linear regression of seven data points gave a correlation coefficient of 0.9996 and an activation energy $\frac{E_A}{R}$ of 4640°K. It can be shown that the intercept of this line is useful as well, assuming n^{th} order Arrhenius kinetics. Integration of Equation 2 for constant temperature gives, for $t = t_{gel}$ at $C = C_{gel}$:

$$\ln t_{gel} = \ln\left|\frac{C_o^{1-n} - C_{gel}^{1-n}}{(1-n)\,A}\right| + \frac{E_A}{RT} \qquad n \neq 1 \tag{5}$$

representing the equation of a line.

To make use of the intercept term of the above to solve for the pre-exponential factor A, the concentration at the gel point C_{gel} was determined by obtaining the conversion calculated by IR absorbance changes at 920 cm^{-1} wavenumbers (12) at the correspond-

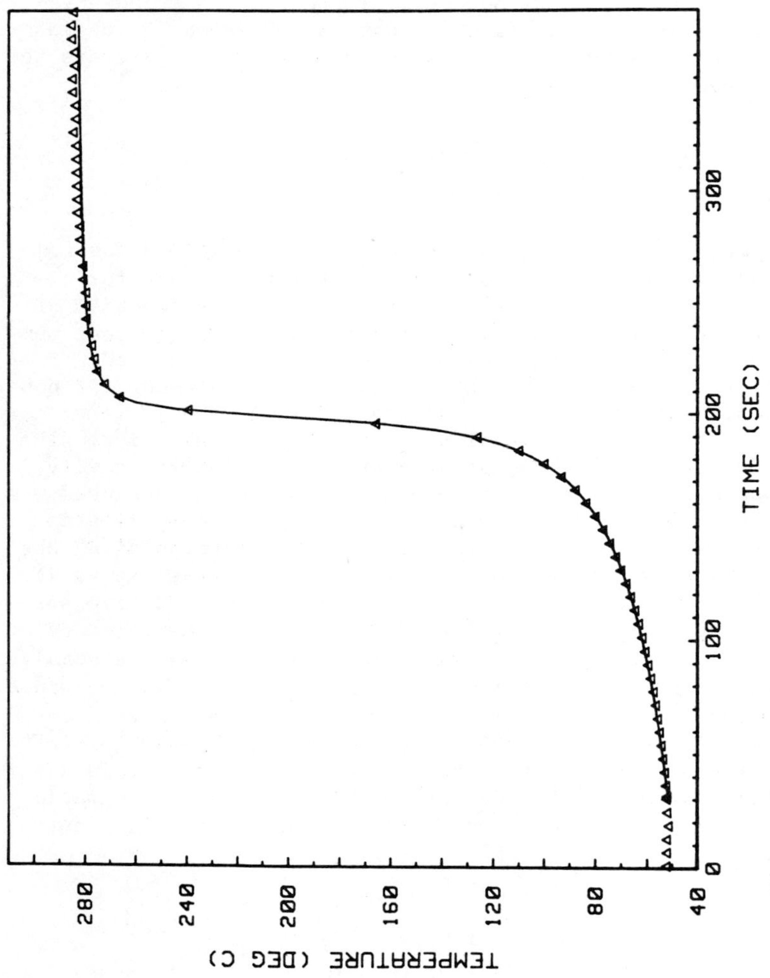

Figure 2. Experimental and correlated adiabatic reaction of epoxy + AEP from T_o = 51.9°C. Δ Experimental, --- correlated. One-sixth the actual number of points taken shown for clarity.

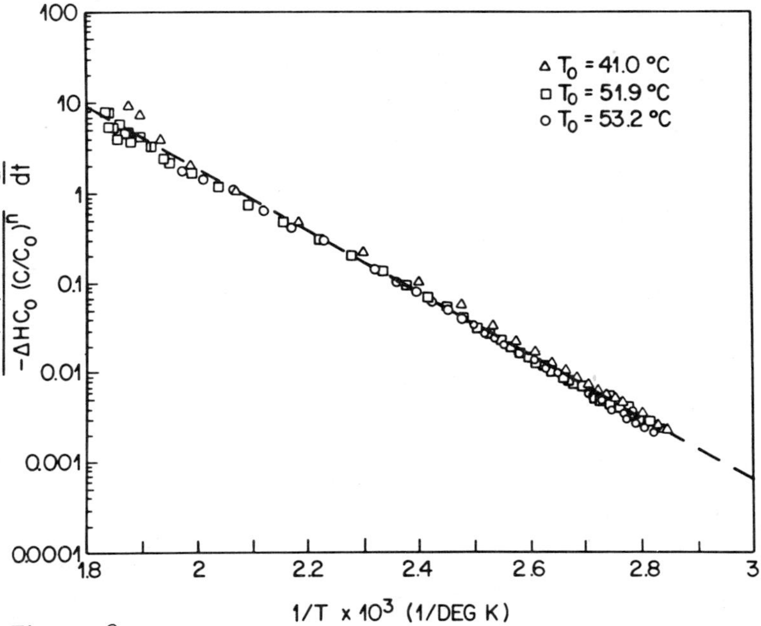

Figure 3.
Data correlation for three epoxy + AEP experiments. ΔT_o = 41.0°C, [ ] T_o = 51.9°C, O T_o = 53.2°C.

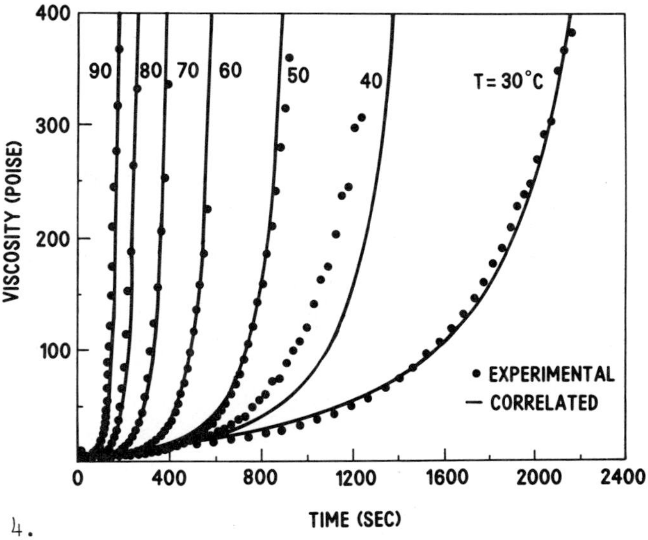

Figure 4.
Isothermal cure viscosity of epoxy, + AEP. Solid curves are described by Equations 6 and 10.

ing temperature and gel time found on the viscometer. A Perkin-Elmer Model 567 spectrometer with heated cell carrier and sodium chloride plates was used to monitor this absorbance peak corresponding to the reduction in the epoxide group. The reactive mixture was mixed at room temperature and placed in a thin film between the two heated NaCl plates whereupon the clock was started. Beer's Law was assumed. Three trials at 40, 60 and 80°C yielded C_{gel} = 2400 equiv/m^3 $\pm$ 7% (X_{gel} = 0.49), and an activation energy in good agreement with that obtained on the viscometer, although with a lower correlation coefficient. A reasonable value of n = 2 was chosen to obtain the low temperature kinetic expression:

$$-\frac{dC}{dt} = .3603 \exp\left|\frac{-4640}{T}\right| C^2 \frac{\text{equiv}}{m^3\text{-sec}} \quad (6)$$

The exact value of the index n is not critical for our purposes and the choice of n = 2 fits the data well. Note the very difference activation energy compared to the high temperature expression (Equation 4). Certainly this discrepancy is beyond any experimental error, indicating the presence of a different curing mechanism at these temperatures. The value of $\frac{E_A}{R}$ = 4640°K compares well with values quoted by others for epoxy-amine systems studied by isothermal gel time measurements at low temperatures (13).

The low and high temperature expressions can then be combined into one equation using empirically determined weighting factors on each term:

$$-\frac{dC}{dt} = 2.704 \exp\left|\frac{-7991}{T}\right| C^{2.80} + .05405 \exp\left|\frac{-4640}{T}\right| C^2 \quad (7)$$

Sample adiabatic reaction paths determined by this equation are shown in Figure 5. Kinetic parameters obtained in this work are summarized in Table I.

Viscosity Correlations. The initial viscosity of DGEBA epoxy + AEP as a function of temperature and at essentially zero conversion can be described by an activated process as well. Linear regression of ln η vs. 1/T to fit the equation

$$\eta = \eta_\infty \exp\left|\frac{E_\eta}{RT}\right| \quad (8)$$

yields the expression

$$\eta = 7.309\text{x}10^{-10} \exp\left|\frac{6960}{T}\right| \text{ poise} \quad (9)$$

with a correlation coefficient of 0.9985.

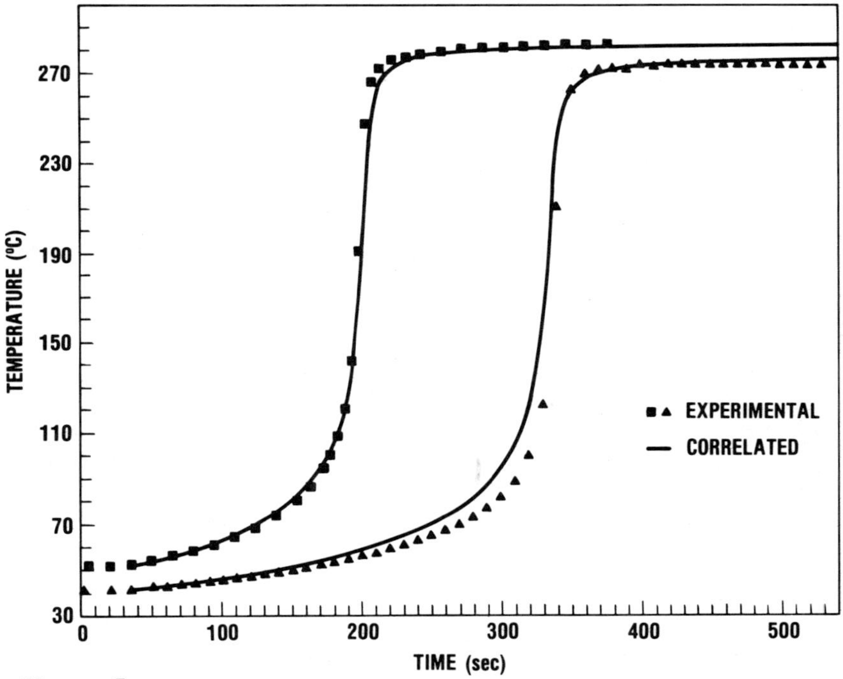

Figure 5.

Two adiabatic reaction paths for epoxy + AEP, at T_o = 41.0°C and T_o = 51.9°C. Solid curves are predictions based on complex kinetics (Equation 7).

For purposes of modeling the filling stage of the RIM molding operation, it is also generally desirable to quantify the increase in viscosity due to polymerization. Therefore, to correlate the viscosity with a developing polymer matrix, a conversion term was added to Equation 8, adapted from Castro and Macosko (14). Since it has been shown that the flow activation energy E_η does not necessarily remain constant with conversion (7), the expression became

$$\eta = \eta_\infty \exp \left|\frac{E^*}{RT}\right| \left|\frac{X_{gel}}{X_{gel} - X}\right| \qquad (10)$$

where E* is a complex function of conversion, and in the present case, temperature as well:

$$\frac{E^*}{R} = \frac{E_\eta}{R} + 2.754\text{x}10^6 \exp\frac{-2400}{T} \quad X \ {}^\circ\text{K} \qquad (11)$$

Note that Equation 10 reduces to Equation 9 for X = 0. Equation 10, describing viscosity as a function of temperature and conversion, is plotted in Figure 6, and, in combination with Equation 6, predicts viscosity and t_{gel} to within a few percent (see Figure 4) over the temperature range explored. Mold filling should not take place outside the range of low conversion and moderate temperature if flow seizure is to be avoided, so these equations are applicable. The discrepancy between the viscometer data and the predicted values at 40°C in Figure 4 is probably due to slight deviation from 40°C on the viscometer during the data acquisition, since the resin cure and viscosity are very temperature sensitive at lower temperatures.

Resin Moldability

Several significant aspects of the RIM operation determine the moldability of a resin in a particular application. These are mixing, mold filling, and resin curing. In addition, the presence of fillers is of importance and can make the difference between a moldable and non-moldable system as the filler affects each of the above steps.

RIM mixing studies have been carried out by Lee, et al. (15), and suggest that the Reynolds number of the fluid through the injection orifices be at least 250 for complete mixing. For the present system this is easily obtainable by adjusting the temperature of the epoxy above 35°C to reduce the viscosity to below 30 poise or so. This is not done without in some way affecting the subsequent filling and curing stages, however, as discussed below. AEP viscosity is nominally 0.2 poise at room temperature and need not be lowered.

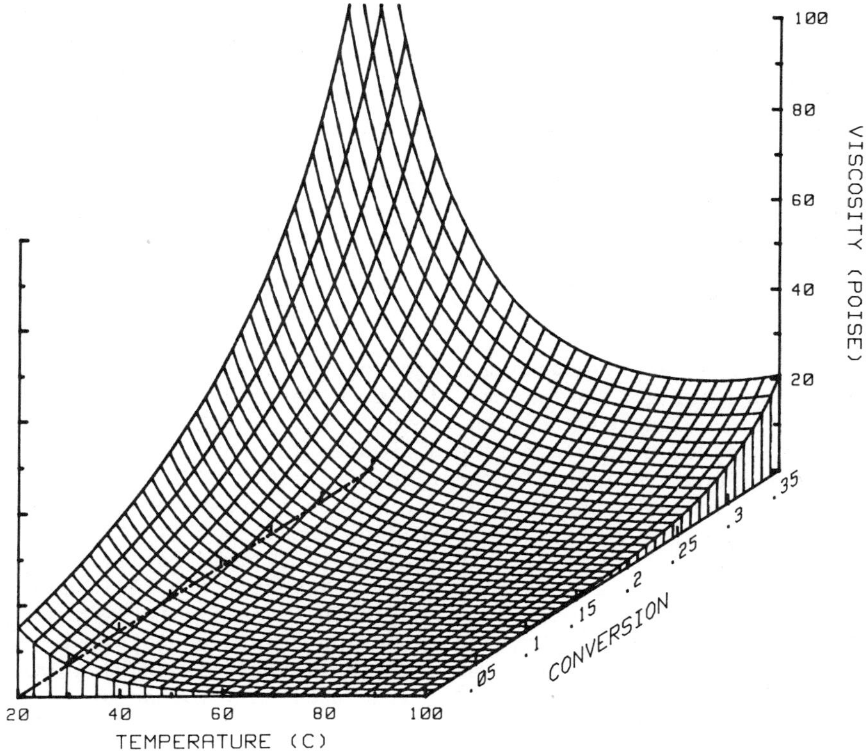

Figure 6.
Viscosity of epoxy + AEP as a function of temperature and conversion (Equation 10).

Mold Filling. Since for very reactive resins the time required for mold filling (8 or more seconds for large parts) may be a significant portion of the adiabatic gel time of the resin, considerable conversion can occur during this step. Indeed, for polyurethanes flow seizure due to gelation is a major concern when establishing processing conditions. Due to the non-uniformity of developing velocity, temperature, and concentration gradients during mold filling, however, it is not a simple matter to predict this event. Furthermore, the non-uniform temperature and concentration profiles that result may have important influences on the subsequent cure. Therefore knowledge of these parameters as a function of location in the mold is valuable in predictive modeling.

A numerical analysis of cavity filling was developed to evaluate and optimize the use of reactive fluids in RIM. This method, which has been previously described in detail (3), employs the marker and cell method for treating transient fluid flows in conjunction with finite difference solutions of the conversion and temperature fields in the pre-polymer and the mold wall. The time-temperature-conversion-viscosity correlations shown earlier for epoxy + AEP were then used in the mold filling simulation.

Results of epoxy modeling indicate that unlike with polyurethanes, however, generally very little conversion occurs during mold filling. This suggests two points: flow seizure due to premature gelation showld not be a problem in epoxy RIM, and an essentially uniform initial condition of $x = 0.0$ will usually suffice for modeling the mold curing sequence under moderate conditions. Often, however, severe molding conditions ($T_W > 140°C$ for this system) are utilized to achieve lower mold residence times. The temperature profiles developed across the fluid due to the filling of the cavity may only then be sufficiently non-uniform to have significant effects on the resulting cure. In such cases, prediction of these profiles for use as the initial condition in the subsequent cure may be necessary for accurate modeling even if the conversion throughout the cavity is negligible.

Shown in Figure 7 are the temperature profiles of epoxy + AEP at the conclusion of a 12 sec fill in a 0.476 cm thick by 12 cm long by infinitely wide cavity, with $T_o = 70°C$ and $T_W = 125°C$. The gate corresponds to $Z/L = 0.0$ in the plot. Minimal conversion has occurred (plot not shown) thus these profiles are almost entirely due to conduction and not heat of reaction. Only mild axial variation of temperature is evident here, and even less would occur with more moderate molding temperatures. Under such circumstances, a one rather than two dimensional solution of the curing stage should suffice, simplifying the modeling procedure.

Mold Curing. Adjusting component and mold temperatures for processing reasons significantly affects the resin curing stage.

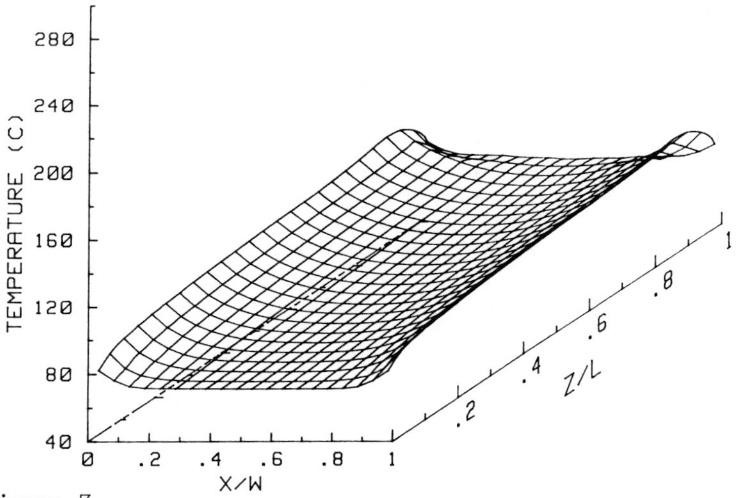

Figure 7.
Temperature profiles at the conclusion of mold filling in a slab geometry with epoxy + AEP. 12 sec fill time, 0.476 section width, 12 cm length, T_o = 70°C, T_w = 125°C.

Heat effects are more important in epoxy RIM than in polyurethane processing because of the usually higher activation energy as well as the much higher heat of reaction. Degradation is rarely if ever encountered in urethane reactions, whereas a 250°C epoxy exotherm can recognizably degrade a part that is nearly adiabatic. Mold and component temperatures as well as part design must therefore be chosen carefully when using epoxies. Since epoxies are generally slower than urethanes, initiative exists for reducing their molding times, but a balance must be maintained in order to minimize cycle time while avoiding thermal degradation.

The numerical treatment of cavity filling can be extended to model curing of the resin to examine heat effects during cure. A one dimensional solution, as mentioned, is often sufficient to predict the temperatures achieved during cure and the time required to demold. Demold time is found by determining the point at which the entire cross section has acheived a minimum ejectable conversion.

Since the highest temperature of a part will usually be reached in the center of its cross section with a highly exothermic system such as epoxy, this is the critical location to monitor the temperature. Predicted and experimentally obtained centerline temperatures for epoxy cured with AEP in 0.476 and 0.635 cm thick cavities are shown in Figures 8 and 9 for a mold temperature of 110°C and a component temperature of 55°C. Experimental data were acquired by mounting a 0.013 cm thermocouple in the center of a 7.5 cm diameter cavity formed by two 0.635 cm thick steel plates separated by an O-ring seal. This unit was placed in a temperature controlled laboratory oil bath and epoxy was injected through a small space in the seal, whereupon temperatures were obtained through the same data acquisition system described previously. Heat conduction during the cavity filling of ≈2 sec was neglected in the model, but a non-isothermal solution to the mold wall temperature during cure was included. Excellent agreement is shown between the model prediction and experiment, both in the shape of the curve and the time and magnitude of the peak. The inflection in the experimental data in the early portion of the curve is probably due to convective flow of the resin in the cavity, but this apparently has no great influence on subsequent temperature profiles. The cure model assumes a static fluid.

Simulations over a wide range of cavity widths and initial component and mold temperatures are possible once we are confident of the model. This exercise will be met with varying degrees of success as the conditions deviate far from those described above due to errors in the kinetic expression and thermal parameters such as the heat transfer coefficient. The trends, however, are correct and can be used to explore all possible molding conditions on the computer in a few hours rather than a few weeks or months in the laboratory. Since the temperature and conversion at any point in the resin are intimately related for a given cure history, a good agreement with temperature data should also pro-

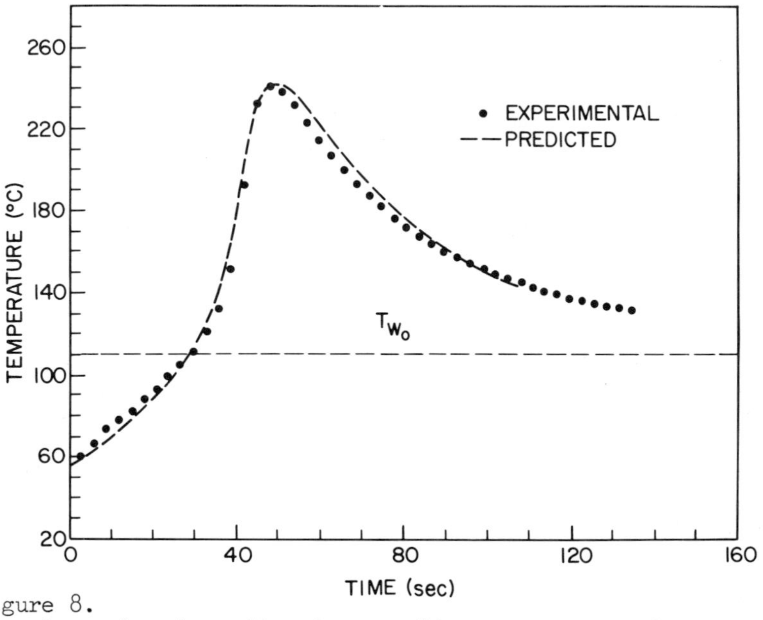

Figure 8.
Experimental and predicted centerline temperature for epoxy + AEP in a 0.476 cm cavity. T_o = 55°C, T_w = 110°C.

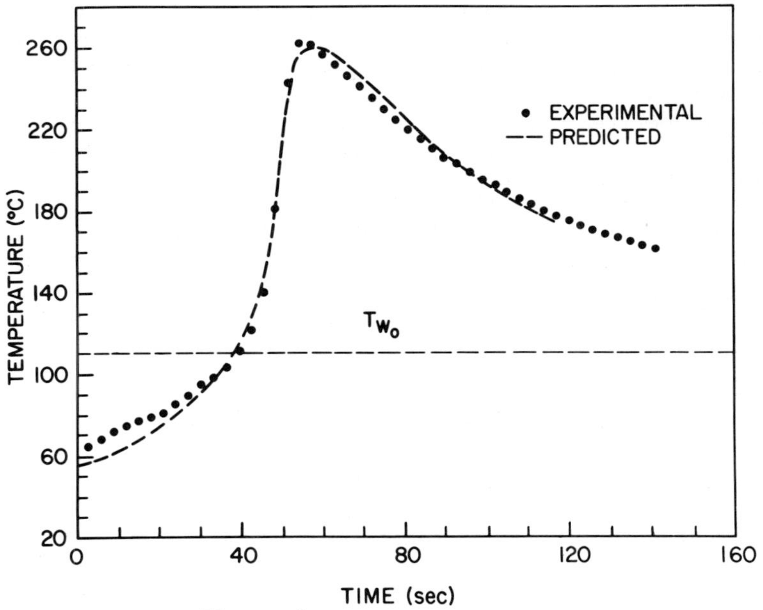

Figure 9.
As above, 0.635 cm cavity.

vide a good agreement with conversion profiles, which are difficult if not impossible to measure experimentally. This can provide the minimum degree of cure in the part at any time, and therefore the time required for ejection. By noting the molding conditions at which the peak temperature reaches the maximum tolerable degradation point (about 265°C in this case), optimal molding conditions for a given mold geometry can be derived. This will generally occur by boosting mold and/or resin temperature to decrease cure time while remaining subject to temperature requirements. For epoxy + AEP in 0.476 cm cavity, a mold temperature of 125°C and a resin temperature of 70°C will provide a demold time of about 45 sec and a peak temperature of 262°C. In a 0.635 cm cavity the same material will process best on the basis of minimum demold time and avoidance of thermal degradation with a mold temperature of 115°C and a resin temperature of 55°C. This results in a minimum demold time of about 75 sec and a maximum temperature of 265°C. The presence of fillers affects the results by increasing heat conduction (usually) and decreasing the effective reaction exotherm.

Summary

To determine the moldability of a prospective reaction injection molding resin, it is useful to empirically characterize the given resin system by time-temperature-conversion-viscosity correlations. It is then possible to choose between different resin systems based on these correlations in conjunction with computer simulations of their behavior and with consideration to process requirements. Epoxy is seen as a viable candidate for RIM, but significant differences with polyurethane exist. Most importantly, the slower reaction rate and higher heat of reaction of epoxies mean that the molding operation is limited by part degradation during cure rather than flow seizure during filling. DGEBA epoxy with AEP as the curing agent is concluded to be a viable off-the-shelf epoxy RIM system with potential cycle times of 2 minutes or less. Specific applications are subject to the physical property requirements, of course, which can be influenced by the addition of fillers, reinforcements, or modifiers.

List of Symbols

A pre-exponential factor, $(\text{equiv/m}^3)^{1-n}\ \text{sec}^{-1}$

C concentration of reactive species, equiv/m^3

C_{gel} concentration at gel point, equiv/m^3

C_o initial concentration of reactive species, equiv/m^3

C_p heat capacity, J/kg-°K

E_A	reaction activation energy, J/equiv
E_η	viscous flow activation energy, J/gm-mol
ΔH	heat of reaction, J/equiv
n	reaction index, dimensionless
r	correlation coefficient, dimensionless
R	universal gas constant, J/equiv-°K
t	time, sec
t_{gel}	time to gel, sec
T	temperature, ° K
T_o	initial temperature, °C
T_w	wall (mold) temperature, °C
T_{w_o}	initial wall (mold) temperature, °C
ΔT_{ad}	adiabatic temperature rise, °K
X	conversion of reactive species $\frac{C_o - C}{C_o}$, dimensionless
X_{gel}	conversion at gel, dimensionless
ρ	density, kg/m^3
η	Newtonian viscosity, poise
η_∞	pre-exponential factor, poise

Literature Cited

1. Domine, J. D.; Gogos, C. G. Polym. Eng. and Sci., 1980, 20, 847.
2. Lee, L. J.; Macosko, C. W. Int. J. of Heat Transfer, 1980, 23, 1479.
3. Manzione, L. T. Polym. Eng. and Sci., 1981, 21, 1234.
4. Kubiak, R. S.; Harper, R. C. SPI Preprints, 35th Annual Technical Conference, 1980, 22-C-1.
5. Levy, N. ACS ORPL Preprints, 1981, 45, 485.
6. Kamal, M. R.; Ryan, M. E. Polym. Eng. and Sci., 1980, 20, 859.
7. Kamal, M. R. Polym. Eng. and Sci., 1974, 14, 231.
8. Kamal, M. R.; Sourour, S. Polym. Eng. and Sci., 1973, 13, 59.

9. Haran, E. N.; Gringras, H.; Katz, D. J. Appl. Polym. Sci., 1965, 9, 3305.
10. Saunders, T. F.; Levy, M. F.; Sevino, J. F. J. Polym. Sci., 1967, 5, 1609.
11. Lipshitz, S. D.; Macosko, C. W. J. Appl. Polym. Sci., 1977, 21, 2029.
12. Lee and Neville, "Handbook of Epoxy Resins"; McGraw-Hill 1967
13. Holtzman, K. A., Bell Laboratories, Murray Hill, N.J.; Manzione, L. T., Bell Laboratories, Murray Hill, N.J.; unpublished results.
14. Castro, J. M.; Macosko, C. W. SPE Preprints, 38th Annual Technical Conference (May, 1980).
15. Lee, L. J.; Ottino, J. M.; Ranz, W. E.; Macosko, C. W. Polym. Eng. and Sci., 1980, 20, 868.

RECEIVED December 2, 1982

15

Water Dispersible Epoxy-g-Acrylic Copolymer for Container Coating

JAMES T. K. WOO, VINCENT TING,[1] JAMES EVANS,[2] CARLOS ORTIZ,[3] GARY CARLSON, and RICHARD MARCINKO

Glidden Coatings & Resins, Dwight P. Joyce Research Center, Strongsville, OH 44136

The role of the interior coatings for cans in contact with food or beverage is demanding indeed. Some of the general requirements for food and beverage can coatings are listed in the following:

1) It must be non-toxic and free from flavors and odors which would adversely affect the quality of the products packed.
2) It must not be affected by the products packed and must prevent reaction between the contents and the metal of the can.
3) It must readily be applicable and rapidly be cured on various types of cans. The product should also be suitable for the type of equipment used.
4) It must have good adhesion and be flexible and tough enough to maintain continuous protective film in subsequent use.
5) It must be economical.

For beer application, special care must be taken with its preservation and it is particularly important with beer cans that coverage should be complete. The critical flavor nature for beer demands the above.

The resins used for the manufacture of food and beverage can lacquers may be either natural or synthetic. Resins combined with drying or semi-drying oils form a class of coatings known as oleoresinous. Other coatings for food and beverage cans are obtained using synthetic resins such as phenolics and epoxies, and acrylic, vinyl, and butadiene polymers. For a brief summary as to what type of resins are used for cans, see Table I.

[1] Current address: IBM Corporation, Boulder, CO 80301
[2] Current address: Glidden Organic Chemicals, Jacksonville, FL 32201
[3] Current address: Tintas Yperanga, Rio de Janeiro, Brazil

0097–6156/83/0221–0283$06.00/0

TABLE I.

3 piece beverage can	Vinyl resin, butadiene polymer.
3 piece sanitary can	Epoxyphenolic, oleoresinous.
2 piece D & I can	Epoxy

All these are solvent based materials, and normally low in solids. With the introduction of pollution legislations, it is imperative that non-polluting polymers be developed for food and beverage applications.

Many approaches have been described in the patent literature. They include water-borne systems[2], powder[3], high solids[4], radiation curable[5], electrocoat[6], etc. The types of resin described include acrylic, polyester, epoxy, vinyl and combinations of these. In this chapter, we will discuss some of the chemistry of a novel approach to interior can coating. This approach involves grafting of carboxyl containing monomers[7] onto high molecular weight epoxy resin. Epoxy resins (epi-bisphenol A type) are known for their adhesion, corrosion resistance and inertness properties. Additionally, they are hydrophobic. Ester grafting of epoxy resins by reaction of carboxylepoxide functionalities has been known for many years[8]. The carboxyl bearing molecule can be an anhydride, fatty acid or acrylic copolymer containing carboxyl functionalities. Upon neutralization with base, the epoxy resin is rendered water dispersible. The obvious weak link in this approach is the presence of ester linkage which will hydrolyze with time, causing instability. In order not to have this weak link, the carboxyl containing monomer was grafted onto high mol. wt. epoxy resin (by carbon-carbon bond formation). The resultant acid containing graft copolymer can be dispersed into water upon neutralization with base. Due to the presence of graft copolymer and absence of ester linkage the emulsion formed has excellent stability.

For most polymers[9], it is often thermodynamically unfavorable for them to form homogeneous mixtures with each other. The key is to produce microheterogeneous polymeric systems so that each component polymer can still retain most of its individual properties while contributing in a synergistic way to provide new macroscopic properties for the materials as a whole.

Often times, polymeric emulsions are formed. All emulsions are thermodynamically unstable because their interfacial area is orders of magnitude greater than the interfacial area of the corresponding coagulated systems. A so-called "stable emulsion" is in reality a meta-stable system. The input of a certain activation energy is necessary for coagulation to occur, and the higher this activation energy, the higher is the meta-stability of the emulsion.

The mechanism of stabilization of polymeric oil-in-oil emulsions consists of the following: 1) Incompatibility of polymers causes the phenomenon of phase separation of polymers in solution, 2) this causes the force which drives the graft copolymer into the interface of polymeric oil-in-oil emulsions, and 3) this causes the formation of coalescence barriers. Without the phenomenon of phase separation which leads to a repulsion of dissimilar polymer chains, there would be no reason for a graft copolymer to accumulate in the interface of an immiscible polymer solution. In order to stabilize a polymeric oil-in-oil emulsion, a graft copolymer must accumulate in the interface of the emulsion and form a protective coating around the emulsion droplets; a so-called coalescence barrier.

In the case of the epoxy-acrylic graft copolymer, the epoxy is not soluble in the monomer mixture, even as low as a 10% solution. However, the epoxy resin being the majority component acts as the continuous phase. The purpose of the graft epoxy-styrene-methacrylic acid copolymer is to lower the barrier at the interface so that a stable oil-in-oil emulsion is first obtained and upon neutralization with a tertiary amine, dimethyl ethanol amine, a stable oil-in-oil emulsion in water is then obtained.

Synthesis and characterization of the graft copolymer has been reported[11]. The Epoxy resin used is a high molecular weight ($\bar{M}n$=10,000) epichlorohydrin-Bisphenol type, and the acrylic is approximaely a 2:1 molar ratio of methacrylic acid/styrene. The Epoxy/acrylic ratio is approximately 80/20 by weight. Grafting is achieved by free radical means.

In free radical grafting[12], there are two possible processes taking place:

(a) Grafting from: the free radical (or other active site) is generated on the backbone and subsequently it initiates the polymerization of monomers to produce branches.

(b) Grafting onto: a growing free radical (or other active species) attacks another preformed polymer preferentially carrying suitable substituents and thereby produces a branch of the preformed backbone.

With a few exceptions[13], most graft copolymers from free radical induced grafting processes usually produce not only the desired graft copolymers, but also homopolymers and other side reactions. Consequently, the exploration and detailed characterization of grafts produced by free radical methods is often cumbersome.

Characterization of the graft copolymer by solvent extraction indicates that a substantial amount of the epoxy resin and of the acrylic are free. About 47% of the epoxy resin is ungrafted, 61%

of the acrylic monomer polymerizes to form free, ungrafted acrylic copolymers. The free epoxy resin isolated is of lower molecular weight than that of the starting epoxy resin and the free acrylic copolymer is higher molecular weight than that of the acrylic copolymer made under the same conditions in the absence of epoxy resin. We were also able to deduce the following:

(I) The amount of grafting between 3% and 15% free radical initiator based on monomer is related to the amount of such initiator used. From fractional precipitation data of graft copolymers prepared with increasing amounts of free radical initiator, there is an increasing level of an acid containing epoxy formed. In fractional precipitation, the epoxy-g-acrylic copolymer mixture was dissolved in N-methyl-pyrrolidone (NMP), a solvent that will dissolve all three components, free epoxy resin, free acrylic copolymer and graft copolymer. Upon adding toluene (a nonsolvent for the acrylic copolymer), the acrylic portion tends to stay in the NMP rich layer and the free epoxy will migrate to the top toluene rich layer, which should have an absence of acid functionalities. However, epoxy resin that has few carboxyl functionalities will behave like free epoxy resin and migrate to the toluene rich layer. With this technique, we were able to determine that samples made with high level of free radical initiator have higher amounts of carboxyl containing epoxy resins. This increased acrylic solubility or relative percent grafting can be shown to be a straight line functions with percent free radical initiator. See Figure I.

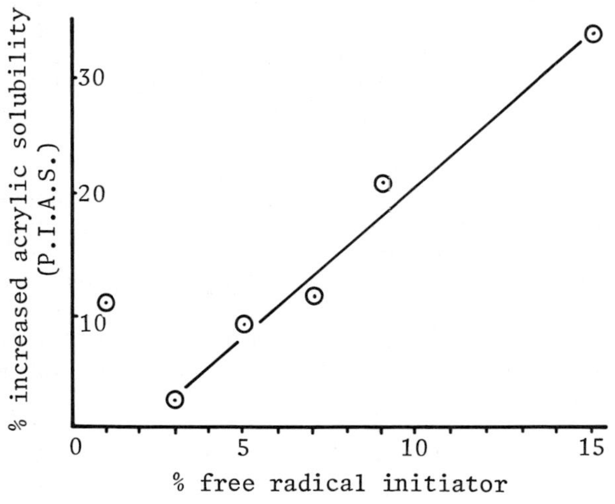

Figure 1. Solubility data of graft copolymers prepared with different levels of free radical initiator.

The minimum amount of free radical initiator used has to be approximately 3% based on monomer weight. Higher levels of free radical initiator seems to give higher level of grafting. Ordinarily, high levels of initiator either cause homopolymerization of monomer (14) or is wasted by recombination and termination processes (15). During radiation grafting (16), a similar phenomenon is observed, that is the percent grafting increases with dosage up to a certain point then levels off. All of our data seems to verify the fractionation results, that is the higher level of free radical initiator produces higher level of grafting.

When the glass transition temperature of the graft copolymers prepared with different amounts of free radical initiator were compared, the data showed a distinct narrowing trend in the glass transition temperature range. This seems to indicate that with higher amounts of free radical initiator, the graft copolymer is becoming more uniform or it may indicate that there may be a higher level of grafting present. See Figure II.

Particle Size Data

The particle size of the aqueous dispersion made with increasing levels of free radical initiator show a significant decrease. The particle size data is summarized in Table II.

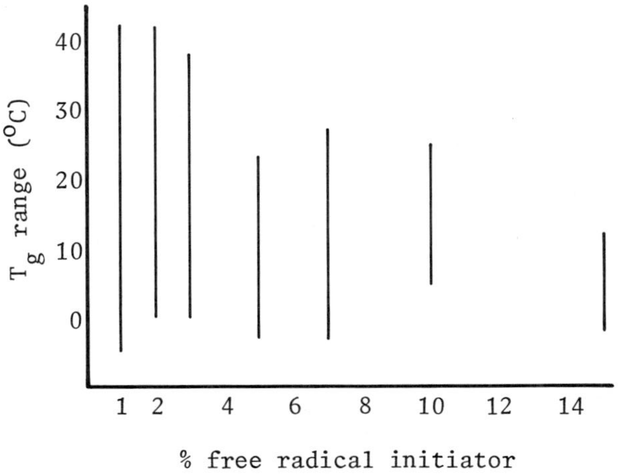

Figure 2. T_g range vs free radical initiator concentration.

TABLE II.

Particle size Distribution of Dispersion from Grafted Copolymers Made With Different Levels of Free Radical Initiator Concentration

% Free Radical Initiator Concentration	Surface Diameter	Volume Diameter	Specific Surface	Wt. Ave. Diameter
1) 1	2.5675	2.7826	18000	2.1953
2) 2	1.6910	1.8393	27030	2.6304
3) 3	0.8716	0.9537	51530	2.8190
4) 5	0.5529	0.5761	94060	1.6759
5) 15	0.3108	0.3158	180400	1.1457

The smaller particle size dispersion made with higher level of free radical initiator concentration suggests that higher levels of graft copolymer are formed with higher levels of free radical initiator. This suggests that the graft copolymer behaves as a surfactant.

(II) Higher molecular weight epoxy resin tends to be preferentially grafted. This is in contrast with grafting on polystyrene(17) where the relative reactivity of polystyrene was found to increase with decreasing molecular weight. This suggests that the reactivity of polystyrene depends mainly on the nature of its coiled conformation in solution. In the epoxy resin case, the molecular weight of the epoxy resin is low compared to that of polystyrene. The higher molecular weight epoxy resin tends to have greater reactivity towards grafting. One of the reasons could be that there are more grafting sites available in a higher molecular weight epoxy resin. This can be demonstrated as follows: In a high molecular weight epoxy resin ($\bar{M}n$ approx. 10,000) there are roughly 34 repeating units of

$$-O-CH_2-\underset{}{\overset{OH}{\overset{|}{C}H}}-CH_2-O$$

in the backbone. There are, therefore, also 34 x 5 ≅ 170 hydrogens that can be abstracted by free radical to form a radical site where initiation of monomers can occur.

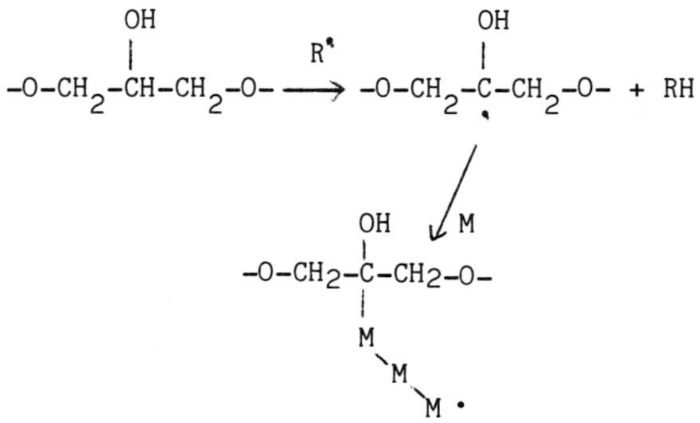

Now if a low molecular weight epoxy resin is used, e.g. $\bar{M}n$ approx. 1000, (Epon 1001 or DER 661) there are about 2

$-O-CH_2-\overset{\overset{OH}{|}}{C}H-CH_2-O-$ in the epoxy backbone, and only 10 abstractable hydrogens.

(III) Epoxy functionalities are not essential for grafting. For a graft copolymer made by the reaction of epoxy with carboxyl functionalities, the presence of epoxy functionalities is essential. Since this approach requires carbon-carbon bond formation, ester linkages are not necessary. Therefore epoxy functionalities can be capped with phenol or benzoic acid, etc., or bisphenol-A terminated epoxy resin can be prepared followed by grafting with acid containing monomers, and resultant epoxy-g-acrylic copolymer can be prepared.

MECHANISM OF GRAFTING

From classical polymerization scheme, there is:

a) Initiation
b) Propagation
c) Chain Transfer
d) Termination

In an idealized case, where transfer mechanism only occurs to a "Foreign Polymer", i.e. transfer to the monomer is negligible, an equation can be derived to show that the rate of grafting is:

$$Vr = k_r \ [Pr.] \ [R] \ [M]$$

Where [Pr.] is the concentration of growing polymer chains formed from the initiator. [R] is the transfer polymer added to the system [M] is the concentration of monomer.

Therefore, the highest yields of grafted copolymer should be obtained under the following conditions:

1) increasing concentration of transfer polymer to a limiting value.
2) high rates of initiation obtained by
   a) increasing initiator concentration
   b) increasing the polymerization temperature

In practice it is not usually possible to take advantage of these conditions since the majority of polymers have limited solubility in "foreign" monomer solutions, particularly at elevated temperatures where the high initiation rates lead to gelation and phase separation. However, substitution of a resin of low molecular weight for the conventional backbone polymer of high molecular weight enables these difficulties to be overcome to a large extent. "Resins" of low molecular weight such as conventional epoxy resins, polyethylene glycols, poly(methoxy acetals), etc. are readily soluble in such monomers as methyl methacrylate, styrene, vinyl acetate, etc., to give high concentrations of "grafting polymer". Even at comparatively high rates of initiation these systems in many cases remain completely compatible and polymerization can be taken to complete conversion without phase separation on a macroscale.

There are numerous examples in the literature [18-21] where free radicals are generated on the polymer backbone followed by grafting of monomer onto the polymer, i.e. initiation of monomer occurring after chain transfer reaction. For example, [19] in grafting monomer onto polyester fiber, active centers seem to be created by direct hydrogen abstraction from the polyester molecules by the primary free radical species benzoyloxy radical or by the secondary free radical species phenyl radical. Benzol peroxide (BPO) was used as initiator. There was no mention of benzoic acid or benzene formation. The other mechanism mentioned was by oxidizing the polyester to hydroperoxide at several points along the chain in a random manner. The hydroperoxide decomposes into the active form at high temperature to produce ultimately macroradicals one of which may be represented as

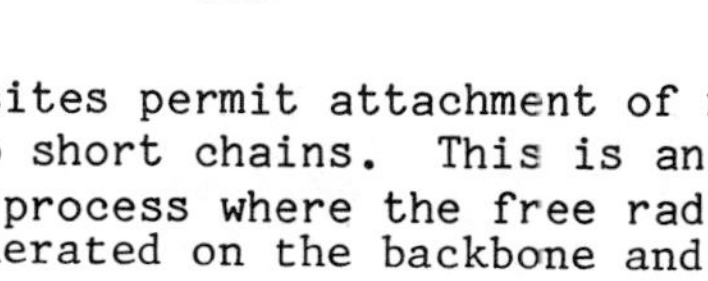

These radical sites permit attachment of monomer molecules which may grow into short chains. This is an example of "grafting from"[12] process where the free radical (or other active site) is generated on the backbone and subsequently it

initiates the polymerization of monomers to produce branches. The other grafting process is the "grafting onto" where a growing free radical (or other active species) attacks another preformed polymer preferentially carrying suitable substituents and thereby produce a branch on the preformed backbone.

Graft epoxy-acrylic copolymer prepared with a free radical initiator is an example of the "grafting from" process. In the case where benzoyl peroxide was used as the free radical initiator, it is determined that about 77% of the free radical initiator instead of causing initiation of monomers, chain transfers with the epoxy resin backbone, followed by the "grafting from" of monomers onto the epoxy resin. Benzoyl peroxide is known[22] to decompose mostly (90%) to the benzoyloxy radical and the (10%) phenyl radical. Mechanisms of grafting can be demonstrated in the following two schemes. (Schemes A and B).

There is probably a very small amount of Scheme B present as the aliphatic protons of the epoxy resin are much more activated towards free radical abstraction than the hydrogen of the 2-butoxy-ethanol-1.

The 23% of the benzoyl peroxide that did not chain transfer would initiate polymerization of monomers to form ungrafted styrene-methacrylic acid copolymer. This could be part of the reason that the ungrafted or free styrene-methacrylic acid copolymer is of higher molecular weight than that of the styrene-methacrylic acid copolymer made under the same conditions in the absence of the epoxy resin. This lower amount of free radical initiator would result in a higher molecular weight copolymer. The other reason is possibly due to viscosity effect, such as Trommsdorf effect where higher molecular polymer is obtained due to lesser chance of termination in a more viscous medium (i.e. in the presence of high molecular weight epoxy resin).

From the data generated so far, an attempt was made to determine the epoxy-acrylic graft copolymer composition.

Seventy-seven percent of the benzoyl peroxide formed benzoic acid by hydrogen abstraction. This was done by using styrene as the only monomer and the resultant benzoic acid titrated is due to abstraction of hydrogen by benzoyloxy radical.

$$11.6 \times 0.77 = 9 \text{ m mole of BPO is involved in grafting}$$

Since about half of the epoxy resin and about two-thirds of the acrylic are free, the grafted composition is now listed in Table (III).

SCHEME A

Transfer with Epoxy resin backbone initially

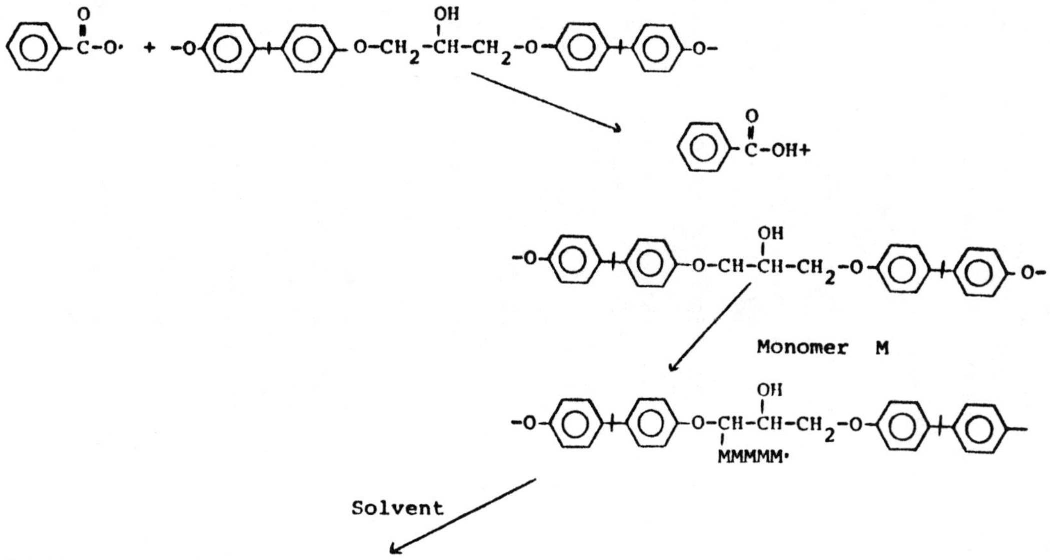

Termination by chain transfer
with solvent or epoxy resin

The latter is probably more likely due to proximity and concentration of the aliphatic hydrogens of the epoxy backbone.

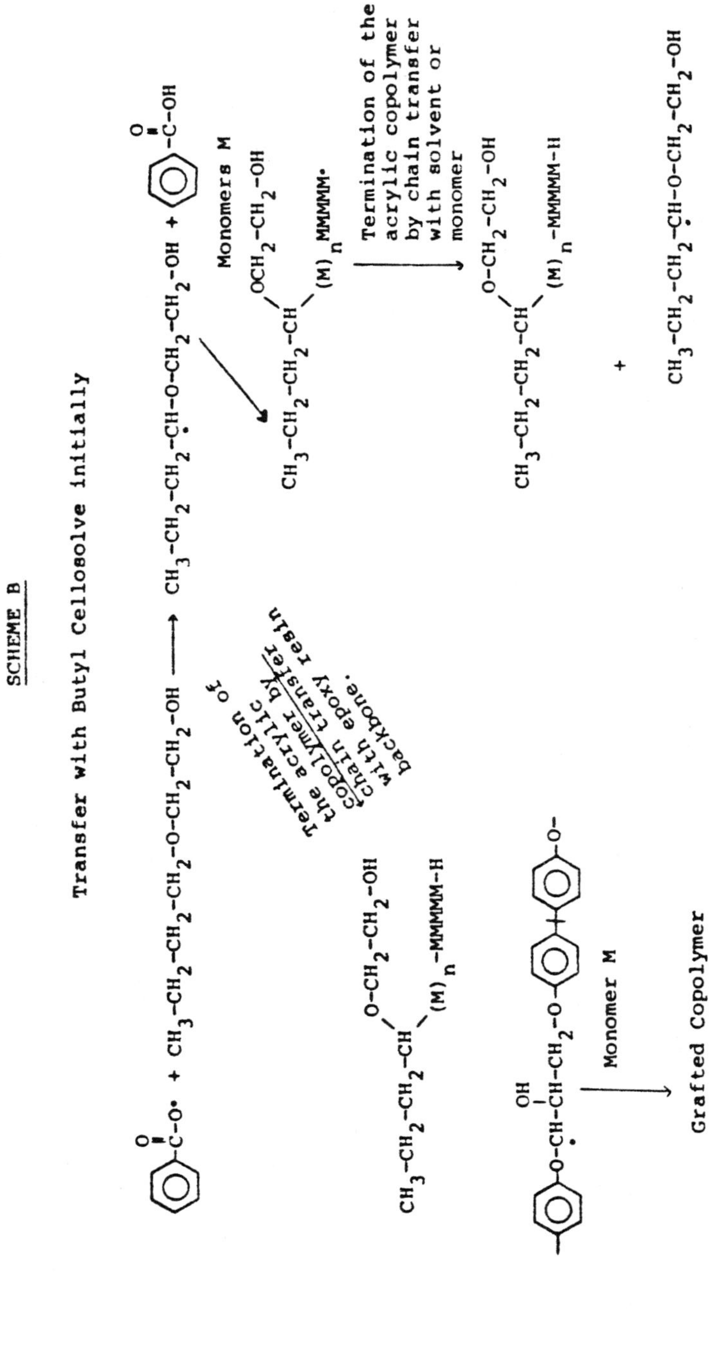
SCHEME B
Transfer with Butyl Cellosolve initially
Monomers M
Termination of the acrylic copolymer by chain transfer with solvent or monomer
Termination of the acrylic copolymer by chain transfer with epoxy resin backbone.
Monomer M
Grafted Copolymer

TABLE III.

Epoxy-Acrylic Graft Copolymer Composition

	Starting Composition m mole	Graft Composition m mole	Ratio
Epoxy	10	5	1
Acrylic	218	87	17
Benzoyl peroxide	11.6	9	2

Therefore, per epoxy molecule, there are about two grafting sites, each grafted chain consists of approx. 9 acrylic units. The composition of the acrylic units are roughly 2:1 methacrylic acid; styrene or

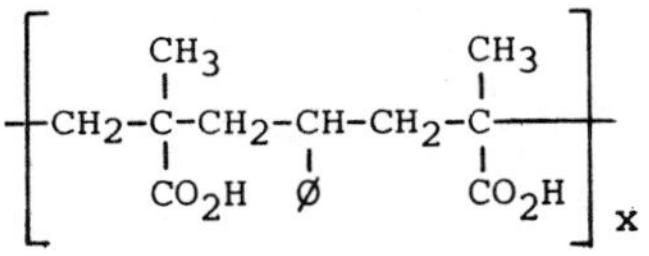

x here is roughly equal to 3.

The composition of the grafted copolymer is thus shown in the following:

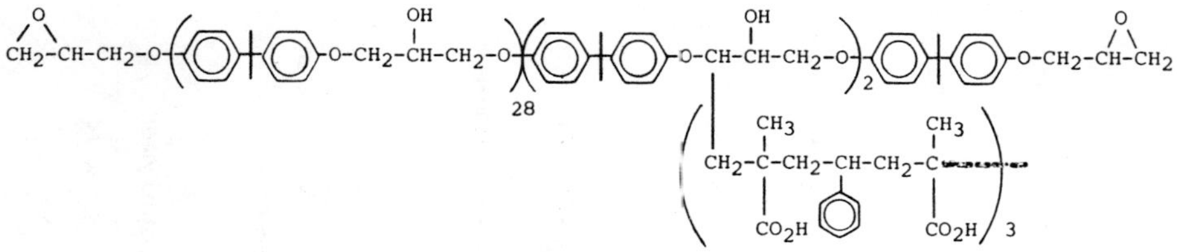

Percent grafting efficiency can be calculated to be:

$$\% \text{ grafting efficiency} = \frac{\text{monomer grafted}}{\text{monomer grafted + free copolymer}}$$

$$= \frac{8}{20} = 40\%$$

The apparent degree of grafting [23] is defined as:

$$G = \frac{W - Wo}{Wo} \times 100$$

G: apparent degree of grafting
Wo: weight of sample before grafting
W: weight of sample after grafting

For the epoxy-acrylic graft copolymer, the apparent degree of grafting is:

$$G = \frac{(\text{free epoxy} + \text{epoxy-g-acrylic copolymer}) - \text{epoxy before grafting}}{\text{epoxy before grafting}} \times 100$$

$$= \frac{88-80 \times 100}{80}$$

$$= \frac{8}{80} = 10\%$$

Anchor/graft ratio based on molecular weight is:

$$A/G = \frac{8858}{828}$$

$$= 10.7$$

<u>Emulsion polymerization using the Epoxy-g-(methacrylic acid-styrene) copolymer as emulsifier.</u>

Emulsion polymerization can be prepared by using different types of emulsifiers. The four most common ones are:

I. Emulsion polymerization in the presence of ionic emulsifiers [24-28] in which radical formation and polymerization initiation occur in aqueous phase and the resulting oligomer radicals precipitate from solution, followed either by homogeneous self-nucleation and formation of initial polymer particles, or by capture of the already existing polymer particles.

Monomer radical formation
initiation in aqueous phase
↓
oligomer radical, grows and precipitates
↙ ↘
homogeneous self-nucleation and formation of initial polymer particles | capture to the already existing polymer particles

This mechanism excludes the possibility of polymer initiation in monomer droplets because of the relatively small global surface compared to that of micelles or of primary polymer particles. However, recently, Ugelstad and Vanderhoff[29] have shown that when monomer dispersion occurs in droplets of very reduced dimensions producing a drastic growth of their surface, the initiation takes place in monomer droplets protected by emulsifiers.

II. Emulsion polymerization in the presence of nonionic emulsifier. In this case, the whole monomer amount is distributed in discrete monomer-polymer particles whose surface is entirely covered by the emulsifier. The number and size of particles is dictated by the emulsifier/monomer ratio, the determining parameter of the process being the ratio between the surface and the volume of droplets.[30-32]

III. The third group includes polymerizations carried out in the presence of nonionic macromolecular emulsifiers, among which ethylene oxide-propylene oxide block copolymers and polyvinyl alcohol are most largely used.

IV. The fourth type includes polymerization in the presence of ionic macromolecular emulsifier. In this case, monomers such as styrene, methyl methacrylate, methacrylic acid, acrylonitrle, etc. were polymerized with the neutralized epoxy-g-(methacrylic acid-styrene) copolymer as emulsifier. The initiator used was a redox system composed of sodium formaldehyde sulfoxylate-tert. butyl hydroperoxide. The base used for neutralization was dimethyl ethanol amine. The molecular weight of the emulsion polymer prepared this way is higher than the epoxy-g-(methacrylic acid-styrene):

	Weight Avg.	Z Avg.
Graft Copolymer	12,600	38,300
Emulsion polymer (with 75 parts graft copolymer and 25 parts monomer)	32,500	409,000

The particle size (determined from a Joyce Loebl Disc Centrifuge Photodensitometer) of the graft copolymer dispersion and that of the emulsion polymer indicate that as polymerization proceeds, there is a development of a large particle size peak which increases in size as polymerization proceeds. The average particle size increases from 0.8 to 1.1µ. It is not clear at this time whether the large particle size peak generated during the polymerization stage is caused by polymerization of monomer swollen in the established droplets or monomer in a separate monomer droplet.

The graft copolymer portion can be as low as 20% of the total system. Surfactants such as Aerosol MA-80, Aerosol OT, Emcol 4600 (sodium Tridecyl Sulfosuccinate), Surfynol 104, Igepal CO-630 (Nonylphenol polyethylene oxide) etc. can be added at the beginning of the polymerization without causing significant difference in the polymerization. Other initiators such as persulfate, benzoylperoxide etc. can also be used.

Emulsion polymerization can be carried out without the presence of acid monomer.

The mechanisms of emulsion polymerization using classical emulsifiers are complex. The use of macromolecular emulsifiers complicates even more the picture of phenomena occurring in emulsion polymerization for the following reasons:

a. Depending on the conditions existing in the medium, the conformation of emulsifier macromolecules in solution will radically change and, hence, accordingly also their emulsifying properties;
b. The radical transfer processes entail the formation of graft polymers which also greatly affect the emulsifying properties, hence, in general, all the emulsion characteristics. For example at high % neutralization, the graft copolymer will be in a highly extended form, at lower % neutralization, the conformation will be in a less extended form. The true nature of the polymeric surfactants present in the graft copolymer is of course not clear. However, one can surmise that the graft copolymer will be a much more efficient surfactant than the free acrylic copolymers of styrene - methacrylic acid.

One of the most important problems of emulsion polymerization is the locus in which the elementary processes (initiation propagation, etc.) take place during polymerization, i.e. monomer droplets, monomer-water interface, monomer-saturated emulsifier micelles, or the monomer aqueous solution. Depending on the existing conditions (emulsifier, monomer and initiator) the polymerization reaction will preponderantly develop in one of the possibilities.

The emulsifying process is probably a dynamic process and the transition between large monomer droplets and those with the smallest dimensions takes place at equilibrium and continuously by a large range of intermediates with different sizes.

Initiation probably takes place preferentially in particles

having the smallest diameter and the largest surface related to the volume and stabilized by the graft copolymer.

SUMMARY

Synthesis of epoxy-acrylic graft copolymer using free radical means was described. Characterization of the graft copolymer by solvent extraction indicated that the graft copolymer is consisted of the following:

1) 47% of the epoxy resin is ungrafted.
2) 61% of the acrylic monomer polymerizes to form acrylic copolymer.
3) 39% of the acrylic monomer is grafted onto 53% of the epoxy resin.

The mechanism of grafting was discussed. The mechanism follows the "grafting from" process in which a free radical is generated on the epoxy backbone and subsequently it initiates the polymerization of monomers to produce branches. The composition of the epoxy-acrylic graft polymer is also discussed. For each epoxy resin molecule, approx. 8,000 molecular weight, there are about two grafting sites. Each grafted acrylic chain consists of approx. 828 molecular weight.

The epoxy-g-acrylic copolymer is an excellent polymeric emulsifier for emulsion polymerization.

Literature Cited

1) S. G. Sheltye, Journal of the Colour Society, July/December, 1974, p. 1-7.

2)

US 3943187	1976	DuPont Co.
US 3997694	1976	DuPont Co.
US 4021396	1977	DuPont Co.
US 4247439	1981	DuPont Co.
US 4255308	1981	DuPont Co.
US 3996182	1976	Whittaker Co.
US 3862914	1975	Mobil Oil
US 4029620	1977	Mobil Oil
US 4028294	1977	Mobil Oil
US 4105614	1978	Mobil Oil
US 4165304	1979	Mobil Oil
US 4179440	1979	Mobil Oil
US 4195005	1980	Mobil Oil
US 4195006	1980	Mobil Oil
US 4303565	1981	Mobil Oil
US 4247659	1981	DeSoto
US 4294737	1981	DeSoto
US 4296011	1981	DeSoto
US 4059550	1977	Celanese
US 4289811	1981	Celanese

	US 4164587	1979	PPG
	US 3960795	1976	PPG
	US 4029621	1977	PPG
	US 3922451	1975	Dow Chemical
	US 4289812	1982	Dow Chemical
	US 3908049	1975	Continental Can
	US 3908050	1975	Continental Can
	US 4094844	1978	Shell Oil Co.
	US 4098744	1978	Shell Oil Co.
	US 4119609	1978	Shell Oil Co.
3)	US 4105613	1978	The Dexter Corp.
	US 4163812	1979	W. R. Grace
4)	US 3960979	1976	DuPont
5)	US 3912670	1975	O'Brien
6)	US 3956091	1976	PPG
7)	US 4212781	1980	SCM Corp.
	US 4285847	1981	SCM Corp.
	US 4308185	1981	SCM Corp.

8) US Patents 3,707,516, 3,943,187, 3,997,694, 4,021,396

9) M. Shen, H. Kawai, AICHE Journal, 24 No. 1, 1 (1978).

10) G. Molau, J. of Polym. Sci., Pt. A, 3, 4235 (1965).

11) J.T.K. Woo, ACS Preprints, Organic Coatings and Plastics Chemistry, 43, p. 142 (1980).
J.T.K. Woo, ACS Preprints, Organic Coatings and Plastics Chemistry, 45, 516, 527, 534 (1981).
J.T.K. Woo, et.al., Journal of Coating Technology 54, #689, p. 41-55, (1982).

12) J. P. Kennedy, Polymer Symposium 64, 117 (1978).

13) G. Mino and S. Kaizerman, J. of Polym. Sci., 31, 242 (1958).

14) R. J. Ceresa, The Chemistry of Polymerization Processes, A.C.S. Monograph #20, 1966, pp. 249-260.

15) D. Sudhabar, K.S.V. Srinivasan, K. Thomas Joseph and M. Santappa, Polymer 22, 491 (1981).

16) S. N. Bhattacharyya and D. Maldas, Journal of Polymer Sci., Polymer Chemistry Edition, 20 939-950 (1982).

17) N. Ohto, E. Niki, Y. Kamiya, J. Chem. Soc. 1976, p. 1416.

18) R. J. Ceresa, J. of Polym. Sci., 53, 1961, p. 9-15.

19) S. H. Abdel-Fattah, S. E. Shalaby, E. A. Allam, A. Habeish, J of Appl. Polymer Sci., 21, 3355 (1977).

20) B. N. Misra, R. Dogra, J. Macromol Sci. Chem., A14(5), p. 763-770 (1980).

21) C. G. Beddows, H. Gil, J. T. Guthrie, Polymer Bulletin, 3, 645-653 (1980).

22) P. W. Allen, G. Ayrey, C. G. Moore, J. of Polym. Sci., 36, 55 (1959).

23) F. Sundardi, J. of Appl. Polym. Sci., 22, 3163 (1978).

24) W. V. Smith and R. W. Ewart, J. Chem Phys., 16, 592 (1948).

25) S. S. Medvedev, Int. Sym. of Macromol. Chem., Pergamon, New York, 1959, p. 174.

26) S. S. Medvedev, Kinetics and Mechanism of Polyreacteoris, Akademiai Kiado, Budapest, 1971, p. 39.
27) R. M. Fitch, Off. Dig. J. Paint Tech. Eng., 37, 32 (1965).
28) R. M. Fitch and C. H. Tsai, Polymer Colloids, Plenum, New York, 1971, p. 73.
29) J. Ugelstad, N. S. El-Aasser and J. W. Vanderhoff, J. Polym. Sci., Polym. Lett., 11, (8), 503 (1973).
30) I. A. Gritskova, S. S. Medvedev and M. F. Margaristova, Koll, Zh., 26, 168 (1964).
31) V. V. Dudukin, I. A. Gritskova, I. N. Medvedeva, S. S. Medvedev, Z. M. Utsinova, N. M. Fodiman, Visokomol. Soedin, 10-A, 456 (1968).
32) S. S. Medvedev, I. A. Gritskova, A. V. Zuikov, L. I. Sedakova and G. D. Berejnoi, J. Macromol. Sci., Chem. A-7 (3), 715 (1973).

RECEIVED December 2, 1982

INDEX

D

E

F

L

M

N

O

P

Jacket design by Kathleen Schaner
Editing and production by Deborah Corson and Paula Bérard

Elements typeset by Service Composition Co., Baltimore, MD
Printed and bound by Maple Press Co., York, PA